The Handbook of Environmental Chemistry

Volume 5 Water Pollution
Part D

O. Hutzinger
Editor-in-Chief

Springer-Verlag Berlin Heidelberg GmbH

Marine Chemistry

Volume Editor: P. J. Wangersky

With contributions by
T. A. Abrajano · S. M. Budge · C. C. Clark
R. J. Helleur · E. D. Hudson · B. D. Johnson
P. E. Kepkay · S. M. Myklestad · C. C. Parrish
K. Pulchan · C. Ramos · V. Svetličić
P. J. Wangersky · R. G. Zika · V. Zitko · V. Žutić

Springer

Environmental chemistry is a rather young and interdisciplinary field of science. Its aim is a complete description of the environment and of transformations occurring on a local or global scale. Environmental chemistry also gives an account of the impact of man's activities on the natural environment by describing observed changes.

"The Handbook of Environmental Chemistry" provides the compilation of today's knowledge. Contributions are written by leading experts with practical experience in their fields. The Handbook will grow with the increase in our scientific understanding and should provide a valuable source not only for scientists, but also for environmental managers and decision makers.

ISSN 1433-6863
ISBN 978-3-642-08531-4

Library of Congress Cataloging-in-Publication Data
The Natural environment and the biogeochemical cycles /
with contributions by P. Craig ... [et al.].
v. <A-F > : ill. ; 25 cm. -- (The Handbook of environmental chemistry :
v. 1) Includes bibliographical refereces and indexes.
ISBN 978-3-642-08531-4 ISBN 978-3-540-48776-0 (eBook)
DOI 10.1007/978-3-540-48776-0
1. Biogeochemical cycles. 2. Environmental chemistry.
I. Craig. P. J., 1944- . II. Series.
QD31. H335 vol. 1 [QH344] 628.5 s

Originally published by Springer-Verlag Berlin Heidelberg New York in 2000
Softcover reprint of the hardcover 1st edition 2000

Production Editor: Christiane Messerschmidt, Rheinau
Cover design: E. Kirchner, Springer-Verlag
Typesetting: Fotosatz-Service Köhler GmbH, Würzburg
Printed on acid-free paper SPIN: 10683826 52/3020me – 5 4 3 2 1 0

Preface

Environmental Chemistry is a relatively young science. Interest in this subject, however, is growing very rapidly and, although no agreement has been reached as yet about the exact content and limits of this interdisciplinary discipline, there appears to be increasing interest in seeing environmental topics which are based on chemistry embodied in this subject. One of the first objectives of Environmental Chemistry must be the study of the environment and of natural chemical processes which occur in the environment. A major purpose of this series on Environmental Chemistry, therefore, is to present a reasonably uniform view of various aspects of the chemistry of the environment and chemical reactions occurring in the environment.

The industrial activities of man have given a new dimension to Environmental Chemistry. We have now synthesized and described over five million chemical compounds and chemical industry produces about hundred and fifty million tons of synthetic chemicals annually. We ship billions of tons of oil per year and through mining operations and other geophysical modifications, large quantities of inorganic and organic materials are released from their natural deposits. Cities and metropolitan areas of up to 15 million inhabitants produce large quantities of waste in relatively small and confined areas. Much of the chemical products and waste products of modern society are released into the environment either during production, storage, transport, use or ultimate disposal. These released materials participate in natural cycles and reactions and frequently lead to interference and disturbance of natural systems.

Environmental Chemistry is concerned with reactions in the environment. It is about distribution and equilibria between environmental compartments. It is about reactions, pathways, thermodynamics and kinetics. An important purpose of this Handbook, is to aid understanding of the basic distribution and chemical reaction processes which occur in the environment.

Laws regulating toxic substances in various countries are designed to assess and control risk of chemicals to man and his environment. Science can contribute in two areas to this assessment; firstly in the area of toxicology and secondly in the area of chemical exposure. The available concentration ("environmental exposure concentration") depends on the fate of chemical compounds in the environment and thus their distribution and reaction behaviour in the environment. One very important contribution of Environmental Chemistry to the above mentioned toxic substances laws is to develop laboratory test methods, or mathematical correlations and models that predict the environmental fate of new chemical compounds. The third purpose of this Handbook is to help in the basic understanding and development of such test methods and models.

The last explicit purpose of the Handbook is to present, in concise form, the most important properties relating to environmental chemistry and hazard assessment for the most important series of chemical compounds.

At the moment three volumes of the Handbook are planned. Volume 1 deals with the natural environment and the biogeochemical cycles therein, including some background information such as energetics and ecology. Volume 2 is concerned with reactions and processes in the environment and deals with physical factors such as transport and adsorption, and chemical, photochemical and biochemical reactions in the environment, as well as some aspects of pharmacokinetics and metabolism within organisms. Volume 3 deals with anthropogenic compounds, their chemical backgrounds, production methods and information about their use, their environmental behaviour, analytical methodology and some important aspects of their toxic effects. The material for volume 1, 2 and 3 was each more than could easily be fitted into a single volume, and for this reason, as well as for the purpose of rapid publication of available manuscripts, all three volumes were divided in the parts A and B. Part A of all three volumes is now being published and the second part of each of these volumes should appear about six months thereafter. Publisher and editor hope to keep materials of the volumes one to three up to date and to extend coverage in the subject areas by publishing further parts in the future. Plans also exist for volumes dealing with different subject matter such as analysis, chemical technology and toxicology, and readers are encouraged to offer suggestions and advice as to future editions of "The Handbook of Environmental Chemistry".

Most chapters in the Handbook are written to a fairly advanced level and should be of interest to the graduate student and practising scientist. I also hope that the subject matter treated will be of interest to people outside chemistry and to scientists in industry as well as government and regulatory bodies. It would be very satisfying for me to see the books used as a basis for developing graduate courses in Environmental Chemistry.

Due to the breadth of the subject matter, it was not easy to edit this Handbook. Specialists had to be found in quite different areas of science who were willing to contribute a chapter within the prescribed schedule. It is with great satisfaction that I thank all 52 authors from 8 countries for their understanding and for devoting their time to this effort. Special thanks are due to Dr. F. Boschke of Springer for his advice and discussions throughout all stages of preparation of the Handbook. Mrs. A. Heinrich of Springer has significantly contributed to the technical development of the book through her conscientious and efficient work. Finally I like to thank my family, students and colleagues for being so patient with me during several critical phases of preparation for the Handbook, and to some colleagues and the secretaries for technical help.

I consider it a privilege to see my chosen subject grow. My interest in Environmental Chemistry dates back to my early college days in Vienna. I received significant impulses during my postdoctoral period at the University of California and my interest slowly developed during my time with the National Research Council of Canada, before I could devote my full time of Environmental Chemistry, here in Amsterdam. I hope this Handbook may help deepen the interest of other scientists in this subject.

Amsterdam, May 1980 *O. Hutzinger*

Seventeen years have now passed since the appearance of the first volumes of the Handbook. Although the basic concept has remained the same some changes and adjustments were necessary.

Some years ago publishers and editor agreed to expand the Handbook by two new open-ended volume series: Air Pollution and Water Pollution. These broad topics could not be fitted easily into the headings of the first three volumes. All five volumes series are integrated through the choice of topics and by a system of cross referencing.

The outline of the Handbook is thus as follows:

1. The Natural Environment and the Biochemical Cycles,
2. Reactions and Processes,
3. Anthropogenic Compounds,
4. Air Pollution,
5. Water Pollution.

Rapid developments in Environmental Chemistry and the increasing breadth of the subject matter covered made it necessary to establish volume-editors. Each subject is not supervised by specialists in their respective fields.

A recent development is the 'Super Index', a subject index covering chapters of all published volumes, which will soon be available via the Springer Homepage http://www.springer.de or http://www.springer-ny.com or http://Link.springer.de.

With books in press and in preparation we have now published well over 30 volumes. Authors, volume-editors and editor-in-chief are rewarded by the broad acçeptance of the 'Handbook' in the scientific community.

May 1997 *Otto Hutzinger*

Contents

Foreword

In this volume we will examine areas of research in marine chemistry which have come into prominance in the past few years, and which are changing the ways in which we look at reactions taking place in seawater. Ever since its beginnings, which I would place as roughly about the time of the Challenger expedition, marine chemistry has developed in two major directions: the first has been its development as a tool for the biological and physical oceanographer, and the second has been its dependence on the laboratory-centered analytical and physical chemists, with their concentration on well-defined, largely inorganic systems.

Until the development of the conductivity method for salinity determination considerable time and effort had to be expended in ensuring the precision and accuracy of the salinity titration; until the automatic analyzers, and more recently, the microchip controllers, nutrient analyses, even if actually performed by biologists or biological technicians, were often under the watchful eye of the resident chemist.

At least until the 1950s, the major interest of most marine chemists seemed to be in the analysis of trace metals, or in measurement of physical constants of the major components, in each case using systems somewhat simpler than the real medium sitting out there in the ocean. This was to be expected; one must walk before one learns to run. However, one must always compare calculated theoretical values with what one finds in nature, and as several researchers had pointed out, the trace metals were largely present in seawater at concentrations well below their solubility limits. Whatever was controlling the concentrations of these elements in the open ocean, it was certainly not likely to be found through pure solution chemistry. There were suggestions in the literature that one should look to reactions with organic materials and with the particulate phase, but in the absence of methods for working with these materials, little progress resulted. With the development of microporous media of defined pore size, the study of heterogeneous reactions has become an important division of marine chemistry, one which needs to be studied in conjunction with the marine biologists.

Another area of research which has become important has been the study of reactions taking place in the surface waters, the surface layers, and the air above the surface, under the input of solar energy. It was commonly thought that the influence of such energy was limited to centimeters; we now know that reactions go on at somewhat greater depths, and that we must also take into account happenings in the air column above the oceans.

Much of what is of interest to the marine chemist, and to the environmental chemist, takes place at interfaces: the sea:air interface, the interface between the surface layer and the seawater, the surfaces of organisms and particulate materials, the sea floor, actually anywhere a discontinuity supplies a place for collection or concentration of materials. Much of what you will find in this volume is concerned with happenings at these interfaces. These are relatively new areas of research; in many of them we are just beginning to understand how careful we must be in order to get trustworthy data.

Even in areas where research has been in progress for many years, we are finding out that in the past we have often not been careful enough to get truly comparable data. Today, in this era of multi-ship international global projects, we can no longer afford not to understand the limits of precision and accuracy of the data we work so hard to obtain. We cannot compare the North Pacific to the South Pacific, or last year's values to the current values, unless we know how good the data really is. We have certainly been guilt in the past of starting major programs with less than satisfactory control on our methods, in spite of the well-known admonition about how one should not begin vast projects. What we eventually do about this problem will probably provide a marker of how serious we are about the science involved.

Finally, we have reviewed, albeit briefly, the current state of our knowledge of man's pollution of the oceans. To do real justice to this topic would require a volume, more likely several volumes, for itself alone. However, this chapter will furnish a way into the topic for the reader who is perhaps more familar with seawater chemistry than with pollution chemistry.

In the fifty years in which I have been associated with the field of marine chemistry, it has expanded almost beyond recognition. Of the equipment I used on my first trip to sea, only the filter photometer would have seemed strange to J. Y. Buchanan, the chemist in residence on the Challenger expedition. As late as the mid-1960s, it was a common dictum that the optimal number of electronic components for a sea-going instrument was less than one. It is now common practice to carry both electronic and computer technicians on cruises, to tend the rooms full of specialized equipment. We are more and more aware that the act of taking the sample can change some of the very properties in which we are interested. The chapters in this volume point to some of the directions we are now taking, and perhaps hint at several more we should be taking.

November 1999 *Peter J. Wangersky*

CHAPTER 1

Marine Organic Photochemistry: From the Sea Surface to Marine Aerosols

Catherine D. Clark, Rod G. Zika

Division of Marine and Atmospheric Chemistry, Rosenstiel School of Marine and Atmospheric Science, University of Miami, 4600 Rickenbacker Causeway, Miami, FL 33149, USA
E-mail: cclark@rsmas.miami.edu, E-mail: rzika@rsmas.miami.edu

Photochemistry in the sunlit surface waters of the ocean is dominated by colored dissolved organic material (CDOM) which lies at the center of a photochemical cycle that critically impacts the marine environment. Sunlight-irradiated CDOM undergoes a complex series of reactions to produce biologically available photodegraded DOM, volatile organic carbon compounds, and reactive species (e.g., excited triplet states, OH, superoxide). These react with each other, trace metals (e.g., iron), and other substances in the ocean in a complex series of reactions that affect marine biota and influence the composition of the surface ocean and marine atmosphere. Although DOM has been the focus of several decades of active research in marine chemistry, fundamental questions about its sources, composition, and reactivity remain. Sea salt aerosols produced at the ocean surface will incorporate some of the photochemically-active organic materials concentrated in the sea surface microlayer, including CDOM. There have been very limited studies on photochemistry in aerosol particles, but likely reactions and yields can be conjectured. The photochemistry of the sea surface microlayer and marine aerosol particles constitutes an important new area of research for marine photochemists.

Keywords: Photochemistry, CDOM, Aerosol, Microlayer.

The Handbook of Environmental Chemistry Vol. 5 Part D
Marine Chemistry
(ed. by P. Wangersky)

1 Introduction

Fifteen years ago, a review article co-authored by one of this chapters authors surveyed the area of photochemistry of natural waters and presented the then current state of knowledge, problems, and future opportunities for research [1]. There has since been a flurry of activity in this area as evidenced by the many reviews and special articles published on organic photochemistry in fresh and marine waters – see, for example, [2–14]. Photochemistry in surface waters is dominated by colored dissolved organic material (CDOM), the most important dissolved light absorbing material in sunlit natural waters. Dissolved organic material (DOM), a predominately humic substance, is the most abundant form of organic matter in the ocean. CDOM refers specifically to the fraction of DOM that absorbs light and hence imparts color to the ocean. Some reviews have given general overviews of aquatic photochemistry [2, 3], but others have dealt more specifically with humic substances [4–8] and the photochemical generation of reactive species from CDOM [9–15]. Great strides forward have recently been made in understanding the source, composition, and contribution of humic substances in surface waters to the global carbon cycle [16–22]. Organic photochemistry has been shown to be critical in the cycling of humic substances, and hence dissolved organic carbon, in sunlit surface waters [16, 19, 20], and many new studies have examined the impact of photochemistry on ecological and bio-geochemical cycles – see, for example, [23, 24].

Many of the photochemically active species present at the sea surface will be transported into the atmosphere in marine aerosols via the sea surface interface. Blough recently reviewed the photochemistry of the sea surface microlayer and the compounds which reside in the thin layer of material at the sea surface [25]. Although very little is known about the optical and photochemical properties of this system, rates of possible photochemical reactions were estimated from measured enrichment factors and known bulk properties. Sea salt aerosols are produced at the ocean surface by the bursting of air bubbles resulting from entrainment of air induced by wind stress [26]. On bursting, these bubbles produce film and jet drops. The enrichment of dissolved organic material in jet drops formed by bubble bursting was first demonstrated over 20 years ago [27]. The newly formed aerosol particles will thus incorporate some of the photochemically-active organic materials concentrated in the sea surface microlayer [28, 29]. These aqueous droplets are then exposed to sunlight in the marine boundary layer above the surface ocean, acting as small efficient photoreactors with a large surface area to volume ratio. Despite the obvious importance of interactions with sunlight in aerosol particles, there have been very limited studies on photochemistry in aerosols. There are two important issues: what are the impacts of photochemical processes on chemistry within the aerosol particles and on the production of reactive gas-phase species?

This chapter begins with a brief introduction to the field of photochemistry and a basic overview of interactions with sunlight in the marine environment. This is followed by a discussion of organic photochemistry in the bulk waters of the surface ocean. This discussion centers on CDOM, since this dominates the photochemistry of the sunlit surface ocean. The reader is referred to several good reviews for a summary of other photoreactions involving trace metals and organometallics [30–34]. The remainder of this chapter focuses on photochemistry in the sea surface microlayer and marine aerosols, and speculates as to likely photochemical reactions and yields. Understanding the photochemistry of heterogeneous environments, such as the sea surface microlayer and marine aerosol particles, will constitute the next frontier for marine photochemists.

2 Photochemistry in the Surface Ocean

2.1 Basic Photochemical Principles

Natural water bodies contain a range of chromophoric substances that absorb sunlight to generate a number of reactive species, and a complex series of secondary photochemical reactions subsequently occurs. Photochemistry is the study of the effect of radiant energy on molecules, and the rates and mechanisms by which reactions proceed when initiated by light. In marine photochemistry the source of radiation is sunlight. This is polychromatic (many colored) with wavelengths ranging from about 290 nm to 800 nm. Photochemical reactions involve changes as diverse as homolytic and heterolytic cleavages, electron transfer (or redox) reactions, and energy transfer reactions [30]. Because oxygen is ubiquitous in natural waters, it generally acts as the most common acceptor of electrons or energy and takes part in a number of secondary reactions.

The fundamental step in all photochemistry is the absorption of light (photons) by molecules. This is the first law of photochemistry, the Grotthus-Draper law – only radiation absorbed by the system is effective in producing chemical changes. Quantifying the absorbance of radiation at a particular wavelength (λ) requires a comparison of the intensity of radiation entering a defined sample geometry ($I_{o\lambda}$) with that leaving it (I_λ). Absorbance (A_λ) is then given by

$$A_\lambda = \log_{10}(I_{o\lambda}/I_\lambda) \tag{1}$$

For absorbing chemicals in a homogeneous medium, the conventional form of the Beer-Lambert Law describes the rate of light absorption over pathlength l as being proportional to the absorptivity of the absorbing substance, where a_λ is a representation of the capacity of the substance to absorb light at a given wavelength:

$$A_\lambda = a_\lambda l \tag{2}$$

When the molar concentration [C] for the absorbing substance is known, the absorptivity is further defined as

$$a_\lambda = \varepsilon_\lambda [C] \tag{3}$$

where ε_λ is the molar absorptivity of the chemical at a fixed wavelength, λ, in units of $M^{-1}\ cm^{-1}$.

For most fresh and marine waters, CDOM dominates the absorption of solar radiation in the high energy UV region. Since this material is too complex to define precisely the chromophores present or the size of the compounds, the molar concentration of the reactant species are not known and cannot be determined. Hence, "molar" absorptivities cannot be defined or used. To deal with this difficult problem, light absorption by CDOM is better expressed by absorptivity (m^{-1}):

$$a_\lambda = A_\lambda \, 2.303/l \qquad (4)$$

where A_λ is the absorbance measured in a spectrophotometer and l is the optical pathlength in m.

Once light energy has been absorbed in the form of photons, the molecule is excited from its initial ground state to a state of higher energy through the transition of an electron to an orbital of higher energy. The excited state molecule will then rapidly undergo photochemical reactions or decay via a number of photophysical pathways to return to its original ground state. Some of the possible photophysical deactivation pathways include non-radiative processes where no light is emitted, specifically internal conversion between singlet states and intersystem crossing between singlet and triplet states. When the ground-state is a singlet, the transition from an excited triplet state is spin-forbidden (disallowed based on quantum mechanical principles) and triplet states will hence be longer lived than the corresponding excited singlet states. In this case, triplet lifetimes are in the range of microseconds to milliseconds compared to nanoseconds for singlets. Triplet states are thought to be responsible for much of the photochemical behavior exhibited by organic materials in natural waters exposed to sunlight. Radiative deactivation pathways involve the emission of light energy as either fluorescence or phosphorescence. Fluorophore is the term given to compounds that emit light following absorption. Fluorescence involves the transition between the lowest vibrational level of an excited singlet state to the ground state singlet and has a short lifetime of the order of nanoseconds. Since phosphorescence involves the transition from the lowest vibrational level of a longer-lived excited triplet state to the ground-state singlet state, it occurs on the longer timescale of microseconds to milliseconds. For more details on fundamental absorption and fluorescence processes than space here allows, the reader is referred to two excellent resources: a book on the kinetics of environmental aquatic photochemistry by Asa Leifer [30] and a review article on photochemical principles applied to DOM by Bill Miller [8].

To approach both photophysical and photochemical processes in a quantitative manner, a relative efficiency, or quantum yield, must be assigned to the process in question. In general terms, this quantum yield (Φ) is given by [8]

$$\Phi = (\text{Number of photoinduced events})/(\text{Number of photons}) \qquad (5)$$

For primary processes, the sum of Φ for all photophysical and photochemical events that occur must equal 1. For a specific photochemical reaction, the quantum yield is more narrowly defined as [8]

$$\Phi_\lambda = \frac{\text{(Moles of product formed or reactant lost)}}{\text{(Moles of photon absorbed)}} \tag{6}$$

The wavelength designation is required since the energy gained by photon absorption depends on the wavelength of the incident light. For photochemical reactions, Φ_λ is very often less than 1 since some portion of the absorbed energy goes to processes other than the photochemical reaction of interest. In photochemical studies, the rate at which a reaction proceeds is another important parameter. For example, with aquatic humic materials, the rate at which low molecular weight carbon products are formed is of great interest. For natural waters where CDOM initiates most photochemical reactions, Miller [8] gives the following relationship for photochemical reaction rates:

$$\text{Rate} = d[P]/dt = \Sigma_\lambda {}_a\Phi_\lambda I_{o\lambda}(1 - 10_\lambda^{-a}) \tag{7}$$

where [P] is the measured molar concentration of the reactant or product, $I_{o\lambda}$ is the average light absorption rate (einsteins L^{-1} s^{-1}), and ${}_a\Phi_\lambda$ is an apparent quantum yield [8].

Just as "molar" absorptivities cannot be defined or used for the complex and poorly-characterized CDOM, neither can a "true" quantum yield be measured. Since the molar concentration of photochemically reactive sites (or chromophores) is not known for CDOM, an apparent quantum yield, ${}_a\Phi_\lambda$, is used in Eq. (7) to reflect the loss of a direct molar relationship between the primary absorber and photochemical result. It is important to remember that any quantum yield measured for photochemical reactions of CDOM is only an apparent quantum yield, observed with respect to the specific conditions of the experiment. For many reactions involving humic substances, the apparent quantum yields usually have high values in the high energy UV-B region (290–320 nm) and approach zero as the wavelength increases in the UV-A/visible region (>320 nm). However, different reactions do have different ${}_a\Phi_\lambda$ spectra, reflecting the complex mixture of chemical functionalities in CDOM [8].

To measure photochemical reaction rates for humic substances in natural waters, one can either measure chemical changes in surface waters in situ and correlate measurements with light intensity data or expose natural waters to light inside a container in a controlled laboratory environment. Detailed descriptions of experimental design approaches and potential artifacts are given in [8, 30]. Although the use of a_λ simplifies the description of light absorption when the chromophore concentration is not known, it is important to remember that the photochemical reaction rates measured for CDOM are very sensitive to irradiation geometry and hence experimental design [8].

2.2 Chromophores in the Marine Environment

Photochemistry in the surface ocean is dominated by CDOM, the most important dissolved light-absorbing substance. Inorganic solutes contribute little, but some transition metals may play a significant role. For example, dissolved

Fe(III) may be photoreduced to Fe(II), with the net effect of accelerated dissolution of its minerals and chelates. The reader is referred to several good reviews for a detailed description of the photoreactions involving trace metals and organometallics [31–34]. This chapter will focus on the organic photochemistry of CDOM.

The basic spectroscopic properties of humic materials have been thoroughly reviewed and will only be briefly discussed here – see, for example, [2, 5, 13, 35]. The absorption of natural waters containing CDOM is characterized by a featureless tail that decreases with increasing wavelength, with essentially no absorption above 550 nm. Since solar radiation at the sea surface begins around 300 nm and extends to the red [36], most solar energy absorbed by CDOM occurs in the UV/visible region between 300 nm and 500 nm, a region with sufficient energy to initiate photochemical processes (Fig. 1). CDOM may also act as a photosensitizer by absorbing solar radiation and transferring this energy to other compounds that cannot absorb sunlight directly, thus inducing secondary "photochemical" reactions [5, 8].

Upon excitation, CDOM fluoresces. This fluorescence is dependent on the excitation wavelength, but in general broad emission bands are observed between 250 nm and 500 nm [37]. The common fluorescent regions observed are depicted in Fig. 2. In sunlight the fluorescence intensity fades more rapidly than the absorbance. This is attributed to photoreactions of DOM that result in bleaching of the fluorescence [38].

Many researchers have studied the intrinsic fluorescent properties of humic substances and DOM as a means of obtaining information relating to structural characteristics and origin [39–49]. Simple excitation and emission spectroscopic measurements provide limited discrimination among various humic fractions, identifying only broad classes of humics and fulvics. Other fluorescence techniques have been applied in attempts to improve differentiation, specifically synchronous fluorescence spectroscopy [43, 47] and total luminescence

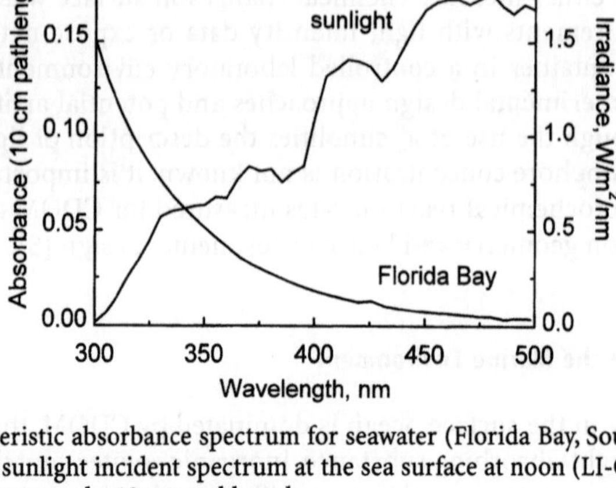

Fig. 1. A characteristic absorbance spectrum for seawater (Florida Bay, South Florida) overlaid by a typical sunlight incident spectrum at the sea surface at noon (LI-COR radiometer). Bulk natural water sample; 10 cm pathlength

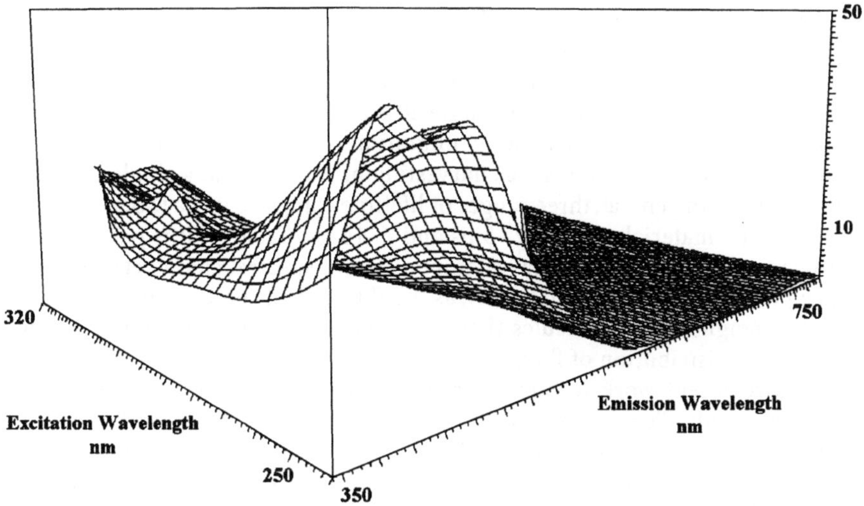

Fig. 2. A three-dimensional excitation/emission matrix for coastal surface seawater (Florida Bay, South Florida) showing excitation/emission maxima at 240/385 nm, 240/435 nm, 320/400 nm and 350/450 nm. Concentrated by ultrafiltration (500 Dalton cut-off membrane); pathlength = 1 mm; Hewlett Packard 1100 FLD (G1321 A)

spectroscopy [37, 48, 49]. Some recent high sensitivity spectrofluorometric studies have shown source-specific differences in fluorescence spectra between freshwater and seawater samples [39, 40], between deep and surface water DOM at the same sites [41], between humic substances of different origins [42, 43], and between different size fractions isolated from natural water samples [44]. Fluorescence measurements have also been used in situ as a measure of DOM concentration profiles [45]. Green and Blough have measured fluorescence spectra for many natural water samples [39, 46], obtaining fairly constant fluorescence quantum yields with 337–355 nm excitation.

Time resolved fluorescence spectroscopy, i.e., lifetime measurement, has several advantages over the steady-state measurements described above. These include insensitivity to variables that may affect fluorescence measurements such as turbidity or scattering in the samples, photobleaching, changes in fluorophore concentration, and optical misalignment. The presence of different lifetimes from samples with similar featureless spectra can provide unambiguous identification of the species. Lifetime measurements are also more sensitive to the fluorophore micro-environment. In spite of these advantages, there have been limited lifetime studies on DOM in seawater [50, 51, 58] and natural freshwater samples of DOM [58].

In one study, the decay of fluorescence for DOM in seawater could be reasonably fitted to a single exponential, with a lifetime of 2 ns [51]. Some recent experiments on DOM from the Baltic Sea have suggested that the fluorescence behavior of humic substances at ambient temperatures may best be described by a mixture of chromophores with only *one* fluorophore, i.e., *one* type of fluorescence molecule with only *one* chemical structure [58]. However, other studies on fulvic acid samples [52–54] and commercial humic acids [55–57] observed multi-exponential decays with fits to three exponentials. A multi-exponential fit would be the more general result in lifetime analyses if one assumes multiple chromophores are responsible for the broad fluorescence bands commonly observed and that energy migration is neither efficient nor uniform for different polymer strands, i.e., harvesting of excitation at a single site does not dominate. In general, three ranges of lifetime components have been obtained for humic materials: <1 ns, 2–5 ns and 7–14 ns (Table 1). It is possible that a single type of fluorophore may acquire different spectral characteristics (and hence possibly fluorescent lifetimes), as the fluorescent centers are surrounded by other neighboring molecules that may influence them to a variable extent, leading to a distribution of fluorophore states in the DOM matrix [58].

In more recent work in our laboratory, fluorescence lifetime studies on a series of South Florida riverine to marine water samples (Shark River in Everglade National Park to Florida Bay in the Gulf of Mexico) also indicated that several fluorophores are present, with multiple lifetimes in the range of 0.8–7 ns [69]. The more marine samples had shorter lifetimes and generally blue-shifted excitation/emission wavelengths compared to the freshwater sam-

Table 1. Fluorescence lifetimes of humic materials and natural waters

Sample	Lifetimes, ns	Excitation/Emission λ, nm	Solution Conditions
Soil-derived fulvic acid [52]	0.53 2.68 8.22	340/480	pH 6.0 0.1 mol l^{-1} KCl 100 mg/l FA
Armdale fulvic acid [53]	~0.2 1.8–3.1 6.5–8.2	355/400	pH 2–7, Ar degassed
Laurentian fulvic acid [54]	0.05 0.43 4.2	355/455	pH 3.5–10
Commercial Aldrich humic acid (3–50 kDa) [56]	0.616 ± 0.048 2.73 ± 0.09 8.48 ± 0.27	340/>420	pH 6.3
Commercial Aldrich humic acid [57]	1.4 7.0	375/465	pH 10,100 mg/l ambient air
Marine waters [50, 51]	1.5–1.9 2.3–2.4	266/460 337/460	pH 8.0; conc. by ultrafiltration
Shark River [69]	0.96 3.72 5.91	337/430 37/430	Bulk natural samples ambient air
Florida Bay [69]	0.66 2.39 6.24	337/430	Bulk natural samples ambient air

ples. This work constitutes the first fluorescence lifetime measurements for DOM in a riverine to marine transition zone.

2.3
Source, Distribution, and Structure of DOM

DOM is the most abundant form of organic matter in the ocean and its influence on physical, chemical, and biological processes in the oceanic carbon cycle are widely recognized [19]. Humic materials, a heterogeneous mixture of complex high-molecular-weight biopolymers, constitute a significant fraction of DOM. CDOM refers specifically to the fraction of DOM that absorbs light and hence imparts color to the ocean. Bulk organic carbon in the ocean is distributed throughout the water column as particulate and dissolved organic material (DOM), with a small fraction in sediments [18], colloids, and the surface microlayer [25, 59]. Average concentrations of dissolved organic carbon or DOC (nominally defined as material passing through filters with a pore-size of 0.45 μm) in seawater are low, at around 80 $\mu mol\,l^{-1}$ in surface waters and ~40 $\mu mol\,l^{-1}$ in deep waters [60]. Methods for collection, extraction, isolation, and analysis of samples have been reviewed [61–63].

Figure 3 summarizes the inputs and cycling of DOM in the sunlit upper layer of the surface ocean, the photic zone. Possible sources of DOM in the ocean include both river inputs and in situ production by phytoplankton. DOM in the marine environment was originally regarded as a complex mixture of both terrestrial material and marine-derived humic substances from phytoplankton

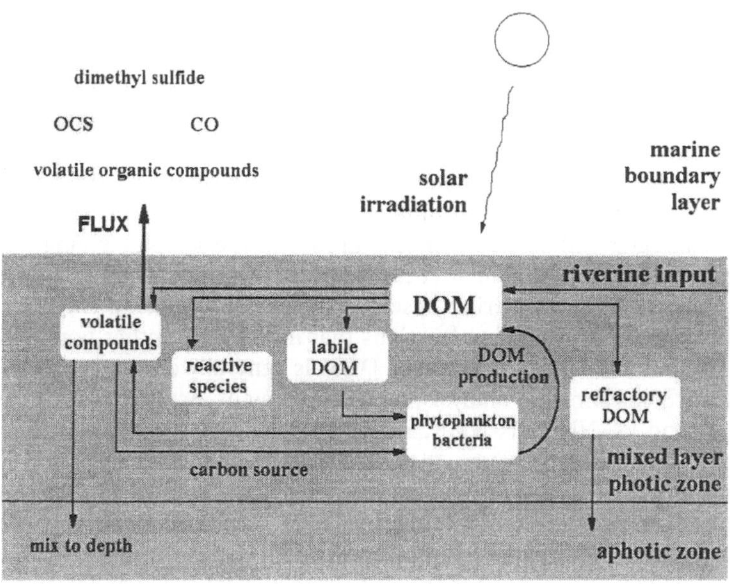

Fig. 3. Sources and cycling of DOM in the surface ocean

[2]. More recent evidence indicates that most DOM in the ocean is of marine origin, since terrestrial inputs only account for about 1% of the total annual input [20]. The old notion that humic matter is simply derived from the degradation of lignin from plants has gradually been replaced by the recognition that these materials may also be formed by microbial activity, where simple monomers from degraded plant material may polymerize into more refractory macromolecules [21]. In the marine environment, DOM may be formed in surface waters by the free radical cross-linking of unsaturated lipids released into seawater by algal/microbial action [21]. Since the mechanism of DOM formation is likely to affect its chemical composition and reactivity, DOM in rivers may very well behave differently to DOM produced in the surface ocean.

Upon absorption of sunlight, DOM undergoes a complex series of reactions which produce photodegraded or labile DOM, which can be accessed by phytoplankton and bacteria for growth (see Fig. 3). Some portion of DOM is refractory (i.e., cannot be further biologically degraded) and is gradually lost to the deep ocean. The photodegradation of DOM also produces volatile organic carbon compounds which transfer to the marine boundary layer and participate in further reactions that impact the local atmosphere. Other photochemical products include a number of reactive species such as excited triplet states, solvated electrons, organic cation radicals, OH and superoxide radicals. These species in turn react with each other, with trace metals such as iron, and with other substances in the ocean in a complex series of secondary redox and photochemical reactions that affect marine biota and influence the composition of the atmosphere over the ocean. CDOM thus lies at the center of a photochemical cycle that critically impacts the marine environment.

Fundamental questions remain about the source, reactivity, and cycling of DOM in the oceans. Since chemical composition is a major factor affecting reactivity, many of the answers are to be found within the chemical structure of this material. Although characterization studies are obviously challenging due to DOM's complex structure and low concentrations in natural waters, new isolation and characterization approaches are providing structural information and insights into the origin and cycling of this carbon reservoir. There are several approaches to isolating material for study from filtered bulk natural water samples. One approach uses chromatographic techniques based on the hydrophobic character and acidic functional groups of the samples, giving fractions of humic materials with similar chemical properties [66]. Another approach is to separate the humic materials based on molecular size using tangential flow ultrafiltration [67], producing fractions that may contain other non-acidic organic compounds. Once fractionated, DOM is generally characterized by techniques such as elemental analysis, molecular weight measurements, acid-base titration, and ^{13}C-NMR, ^{1}H-NMR, and IR spectroscopy [21]. Efforts to characterize DOM with respect to structure and functionality often rely upon sophisticated methods that are difficult to interpret when applied to such complex mixtures of macromolecules like marine DOM [22]. Chromatographic separations have attempted to provide pure components whose structures can be unambiguously identified, but simply confirm that DOM is a complex and poorly resolvable material [22]. These measurements can only provide information about the bulk properties of the sample, i.e., characteristics to which all species contribute to an average value. In general, only average properties can be measured and average structure and functionalities inferred from these results [21, 22].

In general, humic substances, which constitute a significant fraction of DOM, behave as weakly acidic polyelectrolytes that carry net negative charge at neutral pH. An extensive series of early studies showed that humic materials typically consist of a heterogeneous mixture of complex biopolymers with reported molecular weights ranging from 0.5–1000 kDa [64, 65]. In seawater most DOC is actually of a low molecular weight, with only about 20–40% in the more reactive high molecular weight fraction [20]. More recent measurements of the molecular weight of aquatic humic acids give lower values of 1.5–3 kDa [68]. Another approach to separate humic materials based on molecular size has recently been applied in our laboratory to DOM samples from riverine, coastal, and oceanic regions [69]. Flow Field Flow Fractionation (FFFF) refers to a set of chromatography-like analytical separation techniques carried out in unpacked ribbon-shaped channels [70]. In an FFFF channel, separation depends only on the diffusion coefficient (D) of the sample species, i.e., molecular size. FFFF thus has several advantages compared to other common techniques of DOM fractionation, such as gel permeation chromatography and ultrafiltration, which may suffer from adsorption and charge repulsive effects. This technique allows for fractionation of DOM in a seawater matrix, and resolves macromolecules, particulates, and colloids. Following separation, a variety of optical or other characterization techniques may be used in-line to detect the size-fractionated components in the FFFF eluent. Early characterization of freshwater humic substances by this technique gave molecular weight values of around 1.5 kDa [71, 72]. Preliminary studies in our laboratory on marine samples from South Florida waters also indicate molecular weight values in the low kDa range for coastal marine DOM (Fig. 4).

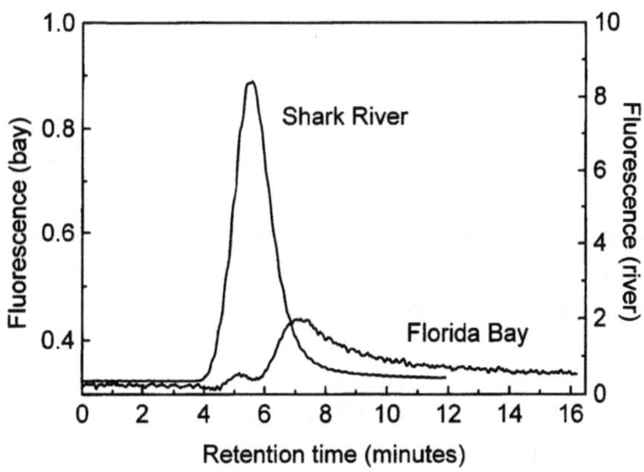

Fig. 4. A comparison of typical fractograms for freshwater vs seawater. Cellulose membrane; Milli-Q water carrier solvent; HP 1100 FLD fluorescence detector ($\lambda_{excitation} = 350$ nm, $\lambda_{emission} = 450$ nm); bulk natural water samples from Shark River and Florida Bay (South Florida)

Elemental analysis of a range of freshwater DOM samples give average compositions of around 30% carbon, 36% hydrogen, 1% nitrogen, and about 25% oxygen [22]. The nitrogen functionalities play an important role in determining properties such as water solubility and metal complexing ability. In recent years, ^{13}C-NMR measurements on DOM have indicated certain common functional groups in humic materials. For example, ^{13}C-NMR studies on a variety of freshwater DOM indicated that the carbon is present on average as about 17% alkyl, 10% aromatic, and 0.05% carboxyl groups [22]. Although the relative abundance of carbon types varies, the same bands are present in ^{13}C-NMR spectra for humic substances from many environments [21]. Differences in relative intensities of the spectral bands show differences in functional group distributions related to differences in source materials and chemical transformations.

Since DOM is present in much lower levels in seawater than freshwater, structural characterization of marine DOM has proven more intractable. However, new isolation techniques such as tangential-flow ultrafiltration isolate relatively large quantities of marine DOM, allowing for more detailed characterization studies [67]. Benner and co-workers have utilized this technique in extensive structural characterization studies of DOM in seawater [73–75], finding that the humic fraction of marine DOM is dominated by aliphatic components with carboxyl functionalities. This is to be expected since oxidative degradation of organic material by microorganisms would result in the addition of carboxylic acid functional groups to the carbon backbones of organic macromolecules [21]. Carbohydrates (specifically polysaccharides), unsubstituted alkyl carbons, and carboxyl/ester carbons are prevalent in all samples [20]. Other recent evidence strongly implicates quinone functionalities in the photochemical activity of DOM in sunlit seawater [4, 76]. In one model of DOM formation incorporating polymerization from reactive monomers, one important class of monomers are the polyphenols or quinones, which are both synthesized by microorganisms and released in lignin degradation [21]. Possible structural characteristic features of marine DOM are outlined in Fig. 5.

One of the most powerful techniques for identifying structural characteristics of complex organic molecules is mass spectroscopy [77]. The technique of pyrolysis-mass spectrometry has been fairly extensively applied to humic substances to investigate structural features – see, for example, [78]. However, there are many problems and limitations associated with this type of measurement due to thermal reactions and system configurations [79]. The recently developed technique of Liquid Chromatography/Mass Spectroscopy^n (LC/MS^n) does not have these shortcomings. Liquid eluents from an FFFF or HPLC system can be directly injected into the mass spectrometer where complex molecules are fragmented, with the resulting fragments repeatedly isolated and re-fragmented to yield compositional information that is difficult to obtain by other means. Using LC/MS^n, structural differences and similarities between DOM fractions and sources can be identified. By drawing correlations between the observed structural features and measured optical characteristics, the identity of the chromophores and fluorophores in DOM can be elucidated. Preliminary studies in our laboratory on DOM from a saline river in Everglades National Park, South Florida, suggest that the chromophoric species (i.e., CDOM) gene-

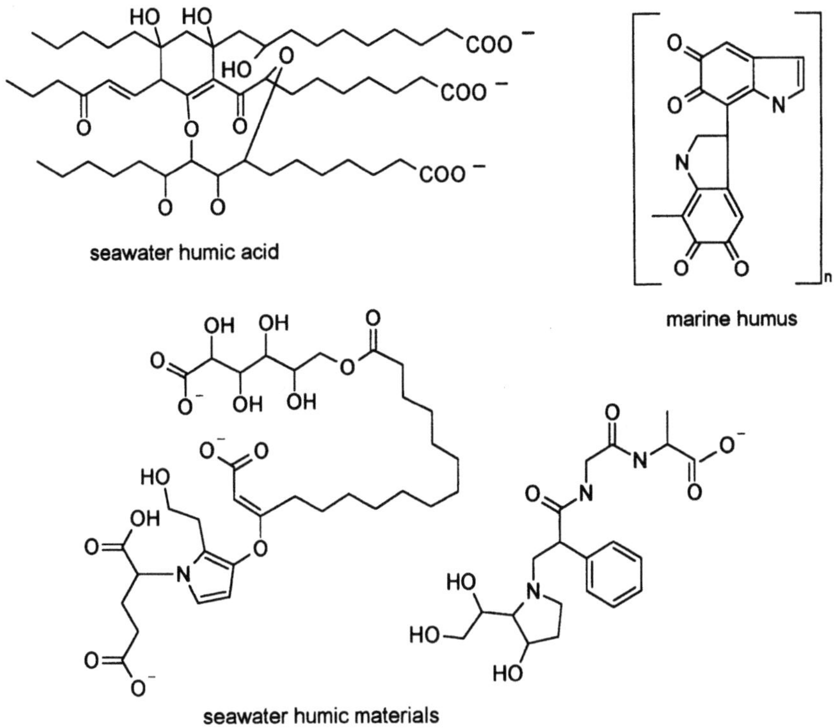

Fig. 5. Possible structural features of marine DOM. Adapted from [1]

rally constitute a smaller fraction of the total DOM mass (Fig. 6). An off-shore coastal sample taken in Florida Bay exhibited a similar relationship, with greater UV absorbance in regions with reduced ion intensities. Preliminary MS/MS and MS^n spectra indicate the presence of multiple carboxyl groups, in agreement with Benner and co-workers' findings that the humic fraction of marine DOM is dominated by aliphatic components with carboxyl functionalities [73–75].

Coupling FFFF or liquid chromatography as the separation technique, with structural characterization by LC/MS^n, is a promising route to elucidating the behavior of DOM in marine systems. This approach may well allow the examination of the optical properties of CDOM on a molecular level by correlating the structural and optical features of sized fractions.

2.4 Photoreactions of CDOM

The absorption of solar radiation by colored DOM (CDOM) leads to the production of short-lived electronically excited states and reactive intermediates [12, 13, 80]. These species participate in a variety of reactions with organic and inorganic substrates that are present in natural waters [1]. The complex inter-

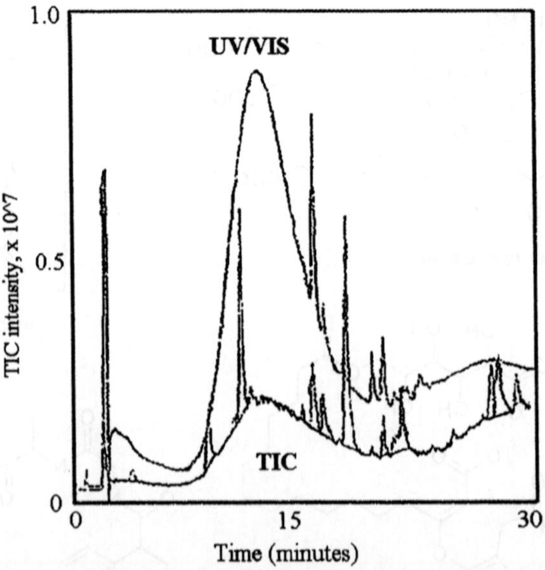

Fig. 6. A comparison of the absorbance of fractions eluting off an HPLC column (UV/VIS) with the total molecular ion intensities of the fractions (TIC). Separation: C-18 reverse phase column with 90:10 v/v water/acetonitrile (both containing 0.1% HAC) to 10:90 v/v over 20 min. Detection: UV absorbance at 254 nm; total ion counts (TIC) with ESI source, m/z scanned 350–2000 in positive ion mode (Esquire LC/MS^n Ion Trap Mass Spectrometer/ Bruker)

play of radical species from irradiated DOM and trace metals was recently highlighted in an article in Chemical and Engineering News [4]. These photodegradation reactions of DOM ultimately lead to the production of sulfur gases, CO, and volatile organic carbon species in sufficient quantities to have some impact on atmospheric composition in certain regions [16, 81–83]. Low molecular weight carbon compounds produced by the photodegradation of DOM stimulate the growth, reproduction, and proliferation of bacteria. Photochemical and biological processes interact in the further degradation of DOM in the marine environment [23].

On irradiation of DOM, the one primary photochemical event is photoionization. This process is less sensitive to environmental factors than triplet formation. All of the remaining photochemistry depends on the subsequent fate of the three primary products of photoionization: triplets, solvated electrons, and cation radicals [13]. Laser flash photolysis studies supplied early evidence for hydrated electron (e^-_{aq}) production on irradiation of DOM, with significant primary quantum yields [9, 12, 84, 85]. The hydrated electron was identified by its characteristic broad absorption spectrum from 700 nm to 750 nm [86, 87]. The formation of e^-_{aq} is thought to result from the photo-ejection of an electron from excited-state humic substances:

$$HS + light \rightarrow HS^* \quad (8)$$

$$HS^* \rightarrow [HS^{+\bullet} + e^-] \tag{9}$$

$$[HS^{+\bullet} + e^-] \leftrightarrow HS^{+\bullet} + e^-_{aq} \tag{10}$$

The primary quantum yield for electron formation, determined on a very short time-scale in laser experiments, is a measure of the electrons initially formed on interaction with light (Eq. 9). The primary quantum yield for electron formation has been measured at 355 nm by laser flash photolysis. Quantum yields were 4.6×10^{-3} to 7.6×10^{-3} for purified humic substances from several different natural waters and 1.7×10^{-3} to 4×10^{-3} for two commercial humic acids (normalized for carbon concentration) [85]. The caged pair generated in Eq. (9) can either collapse back or eject an electron and form the hydrated electron, e^-_{aq}, free in solution (Eq. 10). The steady-state yield, measured with electron scavengers under continuous irradiation, is a measure of the electrons which escape the DOM matrix and are free in solution. The electron thus occurs trapped within the DOM matrix and/or free in solution.

In aerated waters, which are essentially all sunlight illuminated waters, the reactive species present in the highest concentration is O_2 [1], with a concentration at saturation of about 2.5×10^{-4} mol l^{-1}. Hence, O_2 is the dominant scavenger of the photogenerated reactive species. The overall photoprocesses will thus be mediated by reactions with DOM itself and the reactions of the three primary phototransients with ground-state oxygen, $^3O_2$ [13]:

$$^3HS + {}^3O_2 \rightarrow {}^1HS + {}^1O_2 \tag{11}$$

$$e^-_{aq} + {}^3O_2 \rightarrow O_2^- \tag{12}$$

$$R^{+\bullet} + {}^3O_2 \rightarrow RO_2^{+\bullet} \tag{13}$$

where ^3HS is the triplet humic, e^-_{aq} is the hydrated electron, and $R^{+\bullet}$ is the primary radical cation reaction product.

Equation (11) generates singlet oxygen, a reagent for attack at unsaturated centers. $^1O_2$ yields are in the range 0.001–0.026 for natural waters [14, 88]. The organic peroxy radical resulting from Eq. (13) is less well characterized but also quite reactive and potential candidates have been reviewed [12, 13]. Equation (12) produces superoxide ion, which disproportionates rapidly and efficiently to form hydrogen peroxide [12]. Zafiriou and co-workers measured photochemical O_2^- production rates ranging from 0.1–8 nmol l^{-1} min^{-1} in the surface waters of the Eastern Caribbean, accounting for around a third of the total radical production [89]. Zika and Petasne [90] showed that about 60–80% of O_2^- forms H_2O_2. Addition of superoxide dismutase, which catalyzes the disproportionation of O_2^-, increases H_2O_2 yields [90, 91], confirming the involvement of this intermediate in H_2O_2 production. The formation rate of H_2O_2 is related to the concentration of humic substances in water [12]. The photochemical decomposition of H_2O_2 is ~5% of formation rates, indicating that photochemical decomposition is a minor sink for this species compared to biological decomposition [92]. Hydrogen peroxide is thus a secondary product of sunlight induced reactions of CDOM, and may accumulate in surface waters due to its relative stability [10, 93–95].

To measure the quantum yield of e^-_{aq} under photochemical conditions approaching those in natural surface waters, Zepp and co-workers [85] conducted steady-state irradiations of deoxygenated solutions from the Suwanee River and Lake Greifensee in Switzerland. 2-Chloroethanol was used as an electron scavenger, with the quantum yield of Cl^- formed serving as an indicator for e^-_{aq} (the product of Eq. 10). These values are two orders of magnitude lower than the primary quantum yields of 0.005–0.008 measured for Eq. (9), suggesting the role of e^-_{aq} may be minimal compared to other phototransients in reactions with oxygen. In a related study, Cooper and co-workers [14] observed no effects of added electron scavengers on H_2O_2 formation rates in Lake Ontario water. Since the rate constant for the reaction of e^-_{aq} with oxygen is fast and near-diffusion limited ($1.9-2.2 \times 10^{10}$ l mol^{-1} s^{-1} [12]), the electron was hypothesized to remain primarily trapped within the DOM matrix. An alternative triplet mechanism was proposed, where O_2^- (and thus H_2O_2) is formed predominantly from the reaction of O_2 with the excited triplet state of DOM [14]. Lower molecular weight fractions are more efficient in forming $^1O_2$, which is consistent with these fractions having higher quantum yields for intersystem crossing to the triplet state [13]. More recent studies on hydrogen peroxide levels in natural waters attributed the observed correlation of H_2O_2 concentrations with DOM levels to the triplet mechanism for lake waters [96] and e^-_{aq} in seawater [97]. Fujiwara and co-workers [97] generated e^- by irradiating seawater samples at 355 nm and monitoring absorption at 750 nm. A positive correlation was found between the amount of e^- generated and colored DOM in seawater.

Recent work on the role of solvated electrons in intra-DOM reduction processes has demonstrated the importance of "trapped" e^- in reactions with species adsorbed on the DOM matrix [98–100]. Modeling of DOM mediated photoreactions indicated the importance of sorption of molecules to DOM for reaction to occur [98, 99]. This is consistent with the lifetime of e^- precluding escape from the aqueous DOM matrix into bulk solution. Since many important reactions with environmental implications involve binding or adsorption to DOM – see, for example, [3, 101, 102] – the role of matrix effects and the "caged" electron could be very significant. Some workers have suggested that since e^- remains primarily trapped within the DOM matrix, O_2^- must be formed by direct electron transfer from the excited triplet state of DOM to O_2 [14]. However, it is equally if not more plausible that O_2^- may be produced by the reduction of O_2 by radicals or radical ions produced by intramolecular electron transfer reactions from irradiated DOM [25]. The participation of radicals in the production of carbonyl sulfide and carbon monoxide from irradiated DOM in South Florida coastal waters was recently demonstrated by Zika and co-workers [81–83] and potential pathways for the formation of free radicals from irradiated DOM were discussed. Clearly, the relative contribution of e^-_{aq} and associated transients to the photochemistry of DOM has not been unequivocally resolved in the literature.

Anther important photochemical product in surface seawaters is the hydroxyl radical (OH). Zafiriou and co-workers showed that nitrite ions (NO_2^-) in the surface ocean are photolysed by sunlight to produce OH radicals [103, 104]. Nitrate (NO_3^-), an important constituent of seawater, also undergoes photolysis to produce OH [105]:

$$NO_2^- + h\nu \rightarrow NO + O^- \quad (14)$$

$$NO_3^- + h\nu \rightarrow NO_2 + O^- \quad (15)$$

$$O^- + H^+ \rightarrow OH \quad (16)$$

In waters containing high levels of metal ions, OH may also be formed through ligand-to-metal charge transfer reactions and photo-Fenton chemistry [76]. More recently, Mopper and Zhou showed that the total OH production rate in seawater was larger than that expected from the sum of these processes, indicating a missing source of OH production [106, 107]. This missing source was hypothesized to be from the direct photolysis of colored dissolved organic matter (CDOM). In a more rigorous examination of isolated CDOM from the Suwannee River that had been stripped of nitrate, nitrite, and metal ions using a novel OH measurement technique, Vaughn and Blough also found an OH production pathway that could not be attributed to nitrate/nitrite photolysis or Fenton chemistry [76]. This source was dioxygen-independent and again attributed to direct photolysis of CDOM. Since the excited triplet state of benzoquinone and some substituted benzoquinones can abstract a hydrogen atom from water to generate OH, Vaughn and Blough hypothesize that the quinone groups in CDOM play an important role in the photo-initiated formation of OH [76]. These quinone groups may also be the structural moieties responsible for the observed production of hydrated electrons, hydrogen peroxide and O_2^- from CDOM in sunlit seawater (Fig. 7) [4].

A fuller understanding of the photophysical and photochemical pathways by which reactive states and intermediates are produced from the reaction of DOM

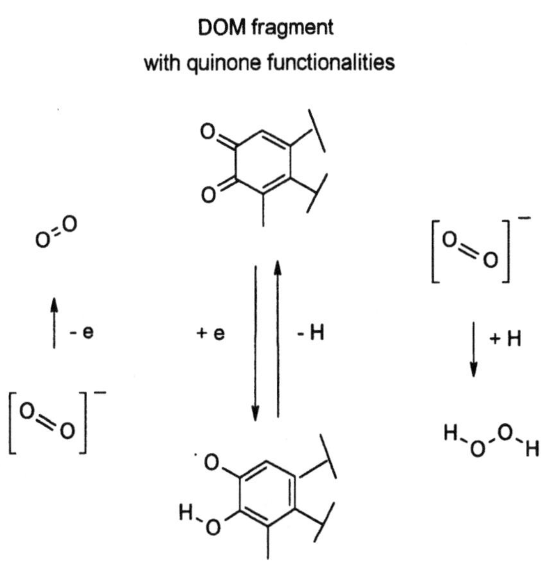

Fig. 7. Possible redox cycling of quinone groups in DOM in sunlit seawater. Adapted from [4]

with sunlight will require further elucidation of the structures of the chromophores and fluorophores responsible for the photochemical activity of CDOM.

3
From the Sea Surface Microlayer to Marine Aerosol

Transients produced from the photochemical reactions of DOM under the action of sunlight in the surface ocean include hydrogen peroxide and OH radicals. These same chemical species are known to play significant roles in atmospheric waters [108]. Hydrogen peroxide plays a crucial role in the oxidation of SO_2 to H_2SO_4, impacting both acid deposition and the global cooling caused by sulfate aerosols, and the hydroxyl radical is a key oxidant in the troposphere. Because of these impacts, it is important to understand the sources of peroxides in atmospheric waters. Models of atmospheric water drop chemistry have focused on gas-to-drop partitioning of peroxides and peroxyl radicals as the sole sources of aqueous-phase peroxides and omitted potential aqueous phase sources of peroxides. However, Faust and co-workers have recently found significant levels of direct aqueous phase photochemical formation of peroxides in atmospheric waters [108–111]. Since many of the photochemically active species present in the surface ocean will be transported into the atmosphere in marine aerosols via the sea surface interface, the direct photochemical production of peroxides and OH radicals from DOM in aerosol particles could play a significant role in marine aerosol chemistry. Despite the obvious importance of interactions with sunlight in aerosol particles, there have been very limited studies on photochemistry in aerosols to date. Photochemistry in the surface ocean has been extensively investigated over the last two decades [2, 6], with recent interest in the sea surface microlayer [25], but no work has been performed to date on sea salt aerosol particles in the marine boundary layer. This section will review the composition of the sea surface microlayer, followed by a brief description of the formation of sea salt aerosol particles at the sea surface. This section concludes with a review of photochemical studies in the microlayer and atmospheric waters, and a discussion of possible photochemical mechanisms, pathways, and products.

3.1
Formation and Composition

Blough recently reviewed the photochemistry of the sea surface microlayer, the thin layer of surface active material at the sea surface, operationally defined as the top 0.03–500 μm depending on the sampling method [25]. The microlayer alters the physical and chemical properties of the air-sea interface. Among the effects of this altered interface are reduced rates of gas exchange, suppression of wind-generated capillary and gravity waves, and changes in the absorption and reflection of light [25]. Materials in the microlayer may undergo more rapid transformations than in bulk seawater due to reactions with radicals and other highly reactive species deposited from the atmosphere [25] and through photochemical processes arising from higher levels of UV radiation at the sea surface

[116]. Perhaps more important, for the purposes of this chapter, the microlayer results in the injection of both surface enriched organic matter and biota into the atmosphere through the collapse of bubbles at the air-sea interface [27, 29].

Although potentially of great importance, the processes occurring within the microlayer have been poorly elucidated to date due both to the inherent difficulties of the research and to the lack of researchers and financial support in this field [25]. For example, very little is known about the optical absorption and emission properties of the microlayer although CDOM is expected to be the dominant chromophore (as for bulk seawater) [25]. In early studies, Carlson showed that the microlayer is enriched relative to bulk waters in material absorbing at 280 nm [112, 113]. This material is most likely composed principally of CDOM [25]. The microlayer may contain up to ten times the concentration of CDOM compared to the bulk water concentration [58]. The current lack of absorption and luminescence spectra of CDOM from the microlayer precludes a more rigorous assessment of differences between bulk and microlayer partitioning of CDOM, e.g., do the more hydrophobic components of CDOM partition selectively into the microlayer? Will the magnitude and spectral dependence of the absorption and emission of light by CDOM in the microlayer vary as for the underlying bulk waters? A wide variety of other inorganic and organic chromophores may also be present at trace levels within bulk seawater and enriched within the microlayer [59]. These include nitrates, nitrites, transition metal complexes, metal colloids, and organic compounds such as flavins and phenols [25]. Although these do not usually contribute significantly to the overall absorption of light, some of these species may still photolyze at appreciable rates.

Sea salt aerosols are produced at the ocean surface by the bursting of air bubbles resulting from entrainment of air induced by wind stress [26]. Surface active films are adsorbed by these bubbles. On bursting, the bubbles produce film and jet drops coated by organic films (Fig. 8). These bubbles are most concentrated in whitecaps associated with wind speeds in excess of 3–4 m s^{-1}. Depending on its size, each bubble can generate as many as ten jet drops with a typical size of 1–2 μm and up to several hundred film drops in the sub-micron range [114]. The enrichment of dissolved organic material in jet drops formed

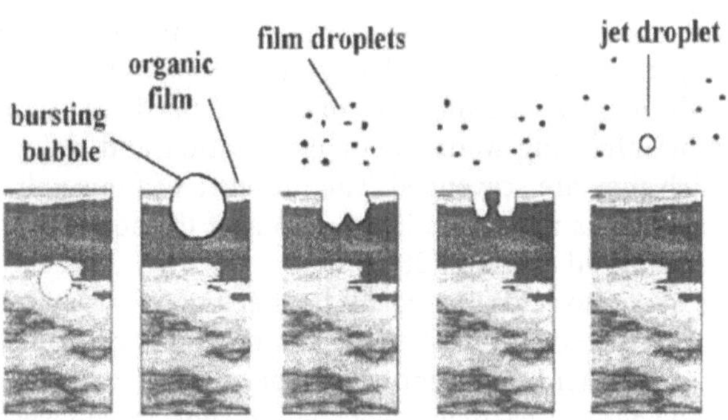

Fig. 8. Formation of marine aerosol particles from bursting bubbles in the ocean, with incorporation of the organic film at the sea surface

by bubble bursting was first demonstrated over 20 years ago [27] and has been more recently examined in the laboratory [29].

The newly-formed aerosol particles will incorporate some of the photochemically-active organic materials concentrated in the sea surface microlayer [28, 29]. These aqueous droplets are then exposed to sunlight in the marine boundary layer above the surface ocean, acting as small efficient photoreactors with a large surface area-to-volume ratio. In the lowest hundred to several hundred meters of the marine boundary layer, sea salt aerosols can dominate the mass concentration and optical properties of the aerosol [114]. Recent evidence indicates that sea salt aerosol particles can also dominate the small size range of the aerosol in the marine boundary layer, comprising a significant fraction of the cloud condensation nuclei [115]. Marine aerosol particles may possibly be transported long distances in storm systems and rained out both locally and remotely. Their constituents may subsequently be transferred to other atmospheric waters like rain, fog droplets, or cloud waters over the ocean and in coastal areas.

Sea salt particles are generally formed in the larger coarse fraction size mode [114]. The composition of sea salt aerosol particles reflects the composition of seawater enriched in organic material (marine-derived sterols, fatty alcohols and acids) that exist in the surface layer [117]. Seawater contains 3.5 wt% sea salt. When first emitted, the sea salt composition is the same as seawater. However, reactions on and in the sea salt particles rapidly modify their chemical composition in the atmosphere. For example, sodium chloride reacts with sulfuric acid from the gas phase to produce sodium sulfate and hydrochloric acid. The latter is then lost to the gas-phase, leading to decreased chloride concentrations in the marine aerosol [117]. In general, atmospheric aerosol particles contain sulfates, nitrates, ammonium, organic material, crustal species, sea salt, hydrogen ions, and water [117]. Nitrate is found in both the coarse and fine size fractions. Sulfate, ammonium, organic and elemental carbon, and certain transition metals are found mainly in the smaller sized fine fraction. Crustal materials and biogenic organics are generally found in the larger sized coarse fraction. Surface active organic molecules are also common constituents of atmospheric aerosol particles, raindrops, and snowflakes in both urban and clean marine regions [28]. Gill and co-workers concluded that films are probably common on atmospheric aerosol particles and may occur under certain conditions on fog droplets, cloud droplets, and snowflakes [28]. If organic films do form, they will increase the lifetimes of aerosol particles and atmospheric water droplets by inhibiting water vapor evaporation and reducing the efficiency with which they are scavenged. The presence of films will also impede the transport of gaseous molecules into and out of the aqueous solution by a factor of several hundred or more [28].

The presence of amino acids in aerosols has been noted for many years [119], but relatively little is known about these or other nitrogen-containing compounds in the atmosphere. A recent review of analyses of marine rainwater shows that marine aerosol particles may contain substantial levels of free and combined amino acids [120], as much as 6.5 $\mu mol\ l^{-1}$ in one study [119]. The most likely source in the remote marine atmosphere is the injection of protein materials into sea salt aerosols during formation at the sea surface [120]. Long range transport from terrestrial sources may also contribute under certain conditions. Amino acids may undergo photodegradation in the atmosphere to simpler species, such as ammonium ions, carboxylic acids, and oxidized forms of sulfur. In one early study, the presence of oxidized compounds, such as methionine, in marine rainwater was attributed to photochemically mediated reactions within sea salt particles [119].

Dissolved organic carbon (DOC) is a major component of the dissolved material in rain in many regions of the world [121]. It includes both biogenic and anthropogenic carbon. Biogenic organic aerosol predominantly consists of lipids and humic material and is often a major fraction of the mass of carbon in aerosol particles and hence in rainwater [118]. Over 40 trace elements are routinely identified in atmospheric particles, along with a wide variety of organic carbon compounds [118]. With few exceptions, dissolved organic carbon is poorly characterized with respect to composition, sources, and seasonal variation. DOC may be a very important reactant in many rainwater reactions such as photochemical processes and trace metal complexation. Average concentrations of DOC measured in rainwater at remote marine sites in the Pacific have ranged from 98 $\mu mol\, l^{-1}$ to 125 $\mu mol\, l^{-1}$ [122, 123]. Coastal rainwater measurements of DOC have been lower at 52 $\mu mol\, l^{-1}$ [124] and 58 $\mu mol\, l^{-1}$ [125]. In one recent study at a coastal site in North Carolina, Willey and co-workers measured volume-weighted average concentrations of DOC of 78 $\mu mol\, l^{-1}$ in hurricane rain systems off the Atlantic Ocean and 23 $\mu mol\, l^{-1}$ in tropical storm marine rain [121]. Different correlation patterns between DOC and other rainwater components in winter and summer continental vs marine rain suggested different sources for the organic material in marine vs continental rain at this site [121]. Organic acids in general are the largest contributor to coastal rainwater DOC at ~40%, mostly in species such as acetic, formic, lactic, oxalic, pyruvic, malonic, maleic, succinic, and other oxoacids [121]. Humic material is the next largest component at 25%, followed by acetaldehyde (5%), formaldehyde (3%), and amino acids [121]. Since marine aerosol particles are transported long distances in storm systems and rained out both locally and remotely, these results suggest that relatively high levels of humic materials exist in marine aerosol particles.

Sea salt aerosol particles form a unique chemical environment, exposed to much higher sunlight photon fluxes than found in the surface waters of the ocean and potentially the site of intense photochemical activity. Species that may interact with sunlight in marine aerosol particles are outlined in Table 2, which lists known constituents and concentration ranges based on and inferred from previous studies. Obviously, these species and levels will vary significantly from study to study, but this merely serves to summarize the most important chromophores commonly found in marine aerosol particles. Based on estimated levels and absorbance cross-sections, specifically molar extinction coefficients and absorptivities, DOM will be the most dominant chromophore in sea salt aerosol particles (as it is in the surface ocean). The photochemical reactions that may occur in the marine aerosol are discussed in more detail in the following section.

Table 2. Potential chromophores in marine aerosol particles: estimated concentrations and relative absorptivities

Substance	Conc., µM	Source	Absorption Peak Wave-lengths, nm	Absorption Coeff., $M^{-1}cm^{-1}$ (350 nm)	Rel. Abs. (at 350 nm)[d]
NO_3^-	0.5–300	seawater [117]	250, 305	<1 [105]	26
NO_2^-	0.034	Sargasso seawater [103]	250, 280, 360	24 [104]	0.07
Fe	5×10^{-3}–3	marine and coastal rainwaters [131]	250, 300	560 [111]	0.24–148
DOC	22–125	marine and coastal rainwaters [121]	250–500 (exponentially decreasing)	~0.01 (in cm^{-1})[a] *conc. dependent*	1000
Amino acids (Tyrosine, Tryptophan)	<0.2–3	marine rainwater [120]	220, 280, 290, 300, 310	<1[b]	0.02–0.26
Pyrene (a PAH)	$0.5–5 \times 10^{-4}$	Boston Harbor seawater [139]	232, 241, 253, 263, 273, 295, 306, 320, 336, 352	0.056[c]	0.0024–0.024

[a] Coastal seawater surface sample from Florida Bay (see Fig. 1); typical DOC ~150 µmol l^{-1}. Absorptivity from Eq. (4).
[b] Creed D (1984) Photochem Photobiol 39:537.
[c] Dabestani R, Ivanov IN (1999) Photochem Photobiol 70:10.
[d] Calculated from Eqs. (3) and (4).

3.2 Photochemistry

Since portions of the sea surface microlayer are injected into aerosol particles via bubble bursting, photochemical studies in this environment may help elucidate processes occurring in the aerosols. Unfortunately, very little is known about the optical and photochemical properties of the sea surface microlayer, although Blough has estimated rates of many possible photochemical reactions from measured enrichment factors and known bulk properties of DOM [25]. Singlet oxygen is formed via energy transfer from the excited triplet states of CDOM to ground state di-oxygen (Eq. 11 in Sect. 2.4). Reactions with $^1O_2$ are probably unimportant in the microlayer, based on estimated steady-state $^1O_2$ concentrations of 10^{-14} mol l^{-1} and the likely residence time of materials in the microlayer. CDOM is the primary source of O_2^- in surface seawaters (Eq. 12 in Sect. 2.4), at a primary production rate of 10^{-11} mol l^{-1} s^{-1} for oligotrophic waters to highs of 10^{-8} mol l^{-1} s^{-1} for coastal waters with higher CDOM concentrations [25]. In the absence of other sinks, the loss of O_2^- is dominated by dismutation to give hydrogen peroxide. In surface waters, the steady-state concentration of

O_2^- could be as high as 400 nmol l^{-1}. However, this concentration may be significantly reduced in the microlayer by other potential sinks. These include reduction by transition metals, reaction with NO to form peroxynitrite, reaction with NO_2, electron transfer and/or coupling reactions with photogenerated phenoxy radicals, and possibly nucleophilic reactions with hydrocarbons in the organic solvent environment of the microlayer [25]. At steady-state and in the absence of other sinks for O_2^-, the rate of H_2O_2 production will be half that of O_2^- [25]. Bacteria or small phytoplankton are the primary sink for hydrogen peroxide. H_2O_2 concentrations in surface waters (~50–200 nmol l^{-1}) do not vary significantly between coastal and marine waters despite large differences in formation rates [91], indicating compensating changes in decomposition rates. OH radicals are formed in seawater from the photolysis of nitrites and nitrates (Eqs. 14–16). The photolysis of H_2O_2 is not a significant source of OH despite the high quantum yield of 0.98 for this reaction, due to the poor overlap between its absorption spectrum and the solar spectrum [25]. The production of O_2^- dominates peroxy radical formation in surface seawaters (Eq. 13), with a rate constant of $\sim 10^9$ M^{-1} s^{-1} and a quantum yield about 2–3 orders of magnitude lower than for nitroxide formation [25]. Different chromophores in DOM are responsible for the formation of different radicals, e.g., acetyl vs methyl radicals. There have been no direct measurements of the production rates and concentrations of any of these species in the microlayer.

The photolysis of CDOM in natural waters produces low molecular weight (LMW) organic compounds, presumably due to radical and fragmentation reactions arising from the net oxidative flow of electrons from CDOM to O_2 [25, 82–83]. The principal products include formaldehyde, acetaldehyde, glyoxal, glyoxalate, pyruvate, acetone, and methylglyoxal [25]. Wavelengths in the UV-B are most effective, with rapid decrease in production rates with increasing wavelength. Quantum yields are of the order of 10^{-6} for formaldehyde [25], low enough for one possible mechanism to be the coupling of two methylperoxy radicals. These LMW organic compounds are taken up by bacteria and respired to CO_2. This coupling between photodegradation of CDOM and uptake of the LMW products by bacteria represents an important route for loss of CDOM in the oceans and an important component of the global carbon cycle [23]. In one recent study, coastal and oceanic microlayer samples were collected with a stainless steel screen along with sub-surface bulk water and analyzed for LMW carbonyl compounds [126]. The enrichment factor in the microlayer vs the subsurface water ranged from 1.2 to 21. A time series at a coastal site showed strong diurnal variations in both the concentrations of the LMW compounds and the enrichment factor, with maxima in the early afternoon and minima in the early morning. The residence times of the LMW compounds in the microlayer were estimated to be of the order of tens of seconds to minutes from a two-layer model, but about one hour based on the observed enrichment factor [126]. Exposure of samples to sunlight gave higher yields of these LMW compounds in the microlayer vs bulk seawater, suggesting that the higher photoproduction rates in the surface microlayer account for the observed enrichment factor. This would correlate with increased CDOM concentrations in the microlayer vs bulk seawater.

In short, although photochemistry in bulk seawater has been extensively studied, very little is known about the microlayer. Even less is known of the photochemical reactions that may be occurring in aerosol particles. Since marine aerosol particles may undergo wet deposition as rain, fog droplets, or cloud waters, photochemical studies of these atmospheric waters over the ocean and in coastal areas can shed some light on potential aerosol processes. Since hydrogen peroxide is formed from the photolysis of DOM and nitrates in aqueous solutions, its presence may serve as a potential marker for the occurrence of these photoreactions within aerosol particles or atmospheric water droplets. Peroxides also play significant roles in oxidation reactions and OH production in atmospheric waters.

In some early studies, Zika and co-workers measured H_2O_2 levels in rainwater collected in South Florida, the Bahamas, Atlantic, and Gulf of Mexico [127, 128]. The H_2O_2 concentration in rainwater is a complex function of gas- and liquid-phase photochemistry, transport dynamics across the gas-liquid interface, and macroscopic transport within clouds. The principal pathway for the formation of H_2O_2 in the troposphere, as opposed to the surface ocean, is via the disproportionation of HO_2 radicals in the gas- or liquid-phase [128]. Measurements of H_2O_2 in rainwater collected at coastal sites in Miami and the Bahamas were in the range $1–7 \times 10^{-5}$ mol l^{-1} [127]. The data suggested that most of the H_2O_2 in precipitation is generated in situ by aqueous phase reactions within cloud water rather than via rainout of gas-phase H_2O_2. Hydrogen peroxide concentrations in marine rain at various sites in the Atlantic were also of the order of $1–8 \times 10^{-5}$ mol l^{-1} on average, about two orders of magnitude higher than levels measured concurrently in the surface ocean [128]. These increased concentrations most likely indicate increased production levels in marine aerosols vs surface waters, or alternate production mechanisms and reduced loss rates. Unpublished data from this laboratory demonstrated the production of H_2O_2 from irradiation of marine aerosols collected on quartz filters, but was not quantified [129]. Three possible mechanisms were proposed to account for this observation: scavenging of gaseous HO_2 by aqueous aerosols, ozone reactions, and photoreactions of DOM injected into sea salt aerosol particles via bubble bursting.

Faust and co-workers have more recently examined the aqueous phase photochemical sources of peroxides (H_2O_2 and organic peroxides) in cloud waters at coastal sites [108–110]. Their research was directed towards answering several key questions [108]:

1. Are peroxides photochemically produced within cloud droplets?
2. What are the rates of this aqueous-phase photochemical peroxide production under typical sunlight conditions?
3. What are the quantum yields for aqueous-phase peroxide production?

Aqueous-phase photochemical peroxide production occurred in >90% of all samples at all sites, with no peroxide production in dark controls [108]. This photoproduction was hypothesized to result from the absorption of light by chromophores in the samples, followed by the formation of peroxyl radical intermediates, the dominant precursor of peroxides in atmospheric waters [108]. Under simulated sunlight, the average rate of total peroxide photoproduction

was of the order of 1 μmol l^{-1} h^{-1}, with individual samples within the range 0–3.0 μmol l^{-1} h^{-1} [108]. There was large variability in production rates between different sites and cloud events [108]. Since other field measurements of peroxides in cloud waters gave levels ≤ 5 μmol l^{-1}, Faust and co-workers concluded that photoproduction in the aqueous phase is a significant, and in some cases dominant, source of peroxides to cloud drops [108]. The average quantum yields for monochromatic illumination at 313 nm and 334 nm were 0.0025 and 0.0013 respectively [108]. Longer-term irradiation over the course of a day suggested that aqueous-phase photoproduction was sustained over many hours at a pseudo-steady state. The initial rate of aqueous-phase production was *linearly* dependent on the actinic flux, whereas the gas-phase production of H_2O_2 (from the disproportionation of HO_2) would show a *squared*-dependence on the actinic flux in most regions of the troposphere [108]. This indicates a seasonal variation of the relative importance of aqueous-phase photochemical sources vs gas-phase sources of peroxides in cloud drops, with in-drop production most important in all seasons except summer [108, 130].

All of the cloud waters examined absorbed UV-VIS light at wavelengths greater than 290 nm, with an exponential decrease of absorbance with increasing wavelength to around 450–500 nm [108]. Although cloud water has not been well-characterized, many chromophoric species, including organic matter, aldehydes, ketones, organic fatty acids, polycyclic aromatic hydrocarbons (PAH), and transition metal complexes are known to be present and could also contribute to the observed UV-VIS absorption [130]. There is a strong similarity between the broad structure-less spectra of marine DOM (see Fig. 1) and the cloud waters, indicating DOM may possibly be a significant contributor of chromophores.

The presence of DOM may be further corroborated by the authors' observation that photochemical production rates correlated with cloud water fluorescence, indicating fluorescent compounds contribute significantly [108]. Singlet oxygen photoproduction was also significant in cloud waters, with a measured concentration of 3×10^{-14} mol l^{-1} to 1.5×10^{-12} mol l^{-1} in sunlit cloud waters [130]. An upper bound of 5×10^{-15} mol l^{-1} can be calculated from the equilibrium partitioning of gas-phase $^1O_2$ [130] into cloud drops, 10–1000 orders of magnitude lower than the actual measured values, indicating in situ aqueous-phase photochemical production [130]. Singlet oxygen is produced from DOM in sunlit waters via the excited triplet state (Eq. 11 in Sect. 2.4) – is this the source in cloud waters? Two important conclusions can be reached from this work:

1. Aqueous-phase photochemical reactions can be an important and dominant source of singlet oxygen and peroxides in cloud drops.
2. Atmospheric models that do not include photochemical sources of peroxides within particles will *seriously* underestimate the concentrations of these oxidants in clouds and fogs [130].

Other light-absorbing compounds that may be present in marine aerosol particles and contribute to photochemistry include transition metals, nitrate anions, amines, polyaromatic hydrocarbons, and other organic molecules. Transition metal complexes (including iron, copper, manganese, nickel) are common constituents of atmospheric droplets, with concentrations ranging over 3–4 orders of magnitude from 10^{-1} μg l^{-1} to 10^{3} μg l^{-1} [131]. In general, the lowest concentrations are found in rain collected at oceanic or coastal sites, indicating low marine inputs compared to terrestrial sources. Fe(III) complexes in acidic solutions absorb light from about 250–350 nm [111], well within the solar spectrum. Early kinetic model studies showed that the photolysis of Fe(III)-hydroxide complexes (Eq. 17) would be a significant source of reactive free radicals in water droplets during the day, even at relatively low concentrations [131]:

$$Fe(OH)^{2+} + h\nu \rightarrow Fe^{2+} + OH \quad (17)$$

At night, Fenton-type reactions would continue to produce free radicals, for example, by

$$Fe(II) + H_2O_2 \rightarrow Fe(III) + OH + OH^- \quad (18)$$

Hydroxyl radicals will react with organic species within atmospheric droplets, as will any hydrogen peroxide that has not been consumed in oxidizing S(IV) to S(VI) [131]. Faust and Hoigne measured quantum efficiencies for the photolysis of $Fe(OH)^{2+}$, the dominant Fe(III)-hydroxy complex between pH 2.5 and pH 5 [111]. Sea salt water droplets rapidly acidify once ejected into the marine boundary layer, and have pH values within this range. Model calculations using the measured quantum yields of 0.14 ± 0.04 at 313 nm and 0.017 ± 0.003 at 360 nm, and absorption spectra, agree with the measured photolysis rate of 6.3×10^{-4} s^{-1} in mid-day June sunlight conditions in Switzerland [111]. More recently, Zuo examined the kinetics of the photochemical cycling of iron coupled with organic acid ligands in cloud and fog droplets [132]. Photolysis rate constants are of the order of $\sim 4 \times 10^{-3}$ s^{-1} for irradiation at 313 nm to $\sim 4 \times 10^{-2}$ s^{-1} for summer noon sunlight irradiation of Fe(III)-oxalate complexes in de-oxygenated solutions [132]. These rates are an order of magnitude higher than those of $Fe(OH)^{2+}$.Oxalic acid, a common constituent in atmospheric waters, was hypothesized to accelerate the sunlight-induced redox cycling of iron by extending the absorption band into the visible region upon complexation, increasing the absorption coefficient and quantum yield of Fe(III) photoreduction. In air-saturated solutions, oxygen further oxidizes the oxalate radical formed in the primary photochemical reaction, generating HO_x radicals and ultimately hydrogen peroxide [132].

It is well-established that nitrate and nitrite anions in aqueous solution absorb sunlight to generate OH radicals via photo-dissociation (Eqs. 14–16 in Sect. 2.4), but this process is rather inefficient compared to the photolysis of iron complexes [131]. A rate constant of 1.9×10^{-7} s^{-1} has been measured for the photolysis of NO_3^- at wavelengths below 330 nm (Eq. 15), and of 3.8×10^{-5} s^{-1} for the photolysis of HNO_2 to NO+OH at wavelengths below 410 nm (Eqs. 14–16) [131]. Other nitrogen-containing chromophores that have been identified in marine aerosol particles are various amino acids [120]. The main chromophores in otherwise "colorless" proteins are the aromatic amino acids phenylalanine, tyrosine, and tryptophan with smaller absorbance from histidine and cystine [120]. These groups absorb above 240 nm, with absorption bands extending out into the solar UV-B range (280–320 nm). However, direct photolysis of amino acids is likely to be a minor process in the environment based on their molar extinction coefficients at the wavelengths of interest [120]. Other possible photochemically-active species in aerosols are organic compounds such as polyaromatic hydrocarbons, produced from the incomplete combustion of organic matter such as fossil fuels and wood. PAHs are compounds, such as pyrene or anthracene, consisting of two or more fused benzene rings in linear, angular, or cluster arrangements [133]. These compounds absorb light in the range from about 260–350 nm, with appreciable molar extinction coefficients of the order of $10^3\ M^{-1}\ cm^{-1}$ [134]. PAH photochemistry has been extensively studied in the gas-phase and urban aerosol particles – see, for example, [135–139]. Ambient concentrations in the gas phase vary from a few ng m^{-3} to values as high as 100 ng m^{-3} reported close to combustion sources and traffic [118]. Coastal waters may contain high PAH levels: in one recent study, pyrene ranged from 10 ng l^{-1} to 100 ng l^{-1} in Boston harbor waters [139]. PAH levels in the clean marine environment should be very low, but aerosol particles in

Table 3. Possible photochemical processes in marine aerosol particles: estimated products and yields

Species	Reaction	Products[a]	Formation Rates, (λ in nm)[b]	Apparent Quantum Yield (λ in nm)[b]	Estimated Relative Yields (350 nm)[c]
NO_3^-	14, 16	NO_2, O^-, OH	$1.9 \times 10^{-7}\ s^{-1}$ (<330) [131]	0.009 (300) [76]	0.6
NO_2^-	15, 16	NO, O^-, OH	$3.8 \times 10^{-5}\ s^{-1}$ (<410) [131]	0.029 (354) [76]	0.005
$Fe(OH)^{2+}$	17	Fe^{2+}, OH	$6.3 \times 10^{-4}\ s^{-1}$ (sun) [111]	0.017 (360) [111]	6
CDOM	8	$^3CDOM^*$	Intersystem crossing	0.01–0.04 [25]	25–100
	11	$(^1O_2)$	? [25]	0.01–0.026 (366) [14, 88]	25–65
	12	(O_2^-)	10^{-11}–$10^{-8}\ Ms^{-1}$ (sun) [25]	? [25]	~3
	from O_2^- [12]	(H_2O_2)	$(4–15) \times 10^{-11}\ Ms^{-1}$ (sun) [25]	0.0006 (350 nm) [12]	1.5
	from quinone*? [76]	$\dot{O}H$	? [76]	~0.002 (300–350 nm) [76]	5

[a] Primary photochemical products (products of secondary reactions in brackets).
[b] Measured in natural waters.
[c] Probable partitioning of a photon on absorption into a marine aerosol particle; estimated from relative absorptivities (Table 2) and apparent quantum yields.
* Photo-excited state.

coastal environments near urban centers may have high enough levels from gas-phase absorption or contaminated waters to play a role in the photochemistry.

Photochemically-active species have been identified in aerosol particles, but the processes following light-absorption are poorly understood. Once a photon is absorbed into a typical marine aerosol droplet, how does it partition? What are the most important likely paths and products? Table 3 summarizes the possible photochemical processes that may occur on light absorption within a marine aerosol particle. Admittedly this is a very approximate estimate based on some typical light-absorbing constituents, their concentrations, relative absorptivities (see Table 2), and known quantum yields and products.

The most likely major photochemical products in marine aerosol particles are OH radicals, singlet oxygen, and hydrogen peroxide, all of which are reactive species that could potentially have some effects on atmospheric processes in the local environment. The purpose of this exercise is not to paint an accurate picture of the photochemistry occurring in marine aerosol particles, a complex and poorly understood environment. Rather it is to provide some possible pathways for future research, and to initiate discussion of the processes initiated by light in this heterogeneous environment.

4 Summary

Photochemistry in the sunlit surface waters of the ocean is dominated by colored dissolved organic material (CDOM), a predominately humic substance that constitutes the most important dissolved light absorbing material in natural waters. The sources, composition, distribution, and photochemical reactions of this material have been the focus of over 15 years of active research in marine chemistry. Many new studies have demonstrated the importance of marine organic photochemistry in ecological and biogeochemical cycles. Upon absorption of sunlight, CDOM undergoes a complex series of reactions which produce labile (biologically available) DOM, volatile organic carbon compounds, and reactive species such as excited triplet states, solvated electrons, organic cation radicals, and OH and superoxide radicals. These species in turn react with each other, with trace metals such as iron, and with other substances in the ocean in a complex series of secondary redox and photochemical reactions that affect marine biota and influence the composition of the atmosphere over the ocean. CDOM thus lies at the center of a photochemical cycle that critically impacts the marine environment. This chapter introduces the field of marine photochemistry with a basic overview of interactions with sunlight in the marine environment. This is followed by a discussion of organic photochemistry in the bulk waters of the surface ocean, specifically the sunlight-initiated reactions of CDOM since this dominates the photochemistry of the photic zone. Although sea salt aerosols produced at the ocean surface by bubble bursting will incorporate some of the photochemically-active organic materials concentrated in the sea surface microlayer, there have been very limited studies on photochemistry in aerosol particles. The remainder of this chapter focused on photochemistry in the sea surface microlayer and marine aerosols, and speculates as to likely photochemical reactions and yields. CDOM may well dominate the absorption of sunlight in marine aerosol particles under certain conditions, with absorption by nitrate ions and iron complexes of secondary importance. The most likely major photochemical products in aerosols are OH radicals, singlet oxygen, and hydrogen peroxide, all of which could potentially have some local effects on atmospheric processes. Understanding the photochemistry of heterogeneous environments, such as the sea surface microlayer and marine aerosol particles, will constitute an important new arena of active research for marine photochemists.

5 References

1. Zafiriou OC, Joussot-Dubien J, Zepp RG, Zika RG (1984) Environ Sci Tech 18:358 A
2. Zika RG, Cooper WJ (eds) (1987) Photochemistry of environmental aquatic systems. ACS Symposium Series 327. American Chemical Society, Washington DC
3. Halmann MM (1996) Photodegradation of water pollutants. Lewis, Boca Raton, p 253
4. Jacoby M (1998) At sea with photochemistry. C&EN, American Chemical Society, Washington DC, October 19
5. Zepp RG (1988) Environmental photoprocesses involving natural organic matter. In: Frimmel FH, Christman RF (eds) Humic substances and their role in the environment. Wiley-Interscience, New York, p 193
6. Miller WL (1994) Recent advances in the photochemistry of natural dissolved organic matter. In: Helz GR, Zepp RG, Crosby DG (eds) Aquatic and surface photochemistry. Lewis, Boca Raton, chap 7
7. Gao H, Zepp RG (1998) Environ Sci Technol 32:2940
8. Miller W (1998) Effects of UV radiation on aquatic humus: photochemical principles and experimental considerations. In: Hessen DO, Tranvik LJ (eds) Aquatic humic substances: ecology and biogeochemistry. Ecological studies, vol 133. Springer, Berlin Heidelberg New York, chap 6
9. Fisher AM, Winterle JS, Mill T (1987) Primary photochemical processes in photolysis mediated by humic substances. In: Zika RG, Cooper WJ (eds) Photochemistry of environmental aquatic systems. ACS Symposium Series 237, American Chemical Society, Washington DC, p 141
10. Cooper WJ, Zika RG, Petasne RG, Plane JM (1988) Environ Sci Tech 22:1156
11. Hoigne J, Faust BC, Haag WR, Scully FE Jr, Zepp RG (1989) Aquatic humic substances as sources and sinks of photochemically produced transient reactants. In: Suffet JH, MacCarthy P (eds) Aquatic humic substance: influences on fate and treatment of pollutants. ACS Symposium Series 219, American Chemical Society, Washington DC, chap 23
12. Cooper WJ, Zika RG, Petasne RG, Fischer AM (1989) Sunlight-induced photochemistry of humic substances in natural waters: major reactive species. In: Suffet JH, MacCarthy P (eds) Aquatic humic substance: influences on fate and treatment of pollutants. ACS Symposium Series 219, American Chemical Society, Washington DC, chap 22
13. Bruccoleri A, Lepore G, Langford CH (1994) The physical chemistry of photochemical oxidant generation in natural water systems. In: Nriagu JO, Simmons MS (eds) Environmental oxidants. Wiley, New York, chap 7
14. Cooper WJ, Shao C, Lean DRS, Gordon AS, Scully FE (1994) Factors affecting distribution of hydrogen peroxide in surface waters. In: Baker LA (ed) Environmental chemistry of lakes and reservoirs. Adv Chem Series 237, American Chemical Society, Washington DC, p 393

15. Blough NV, Zepp RG (1995) Reactive oxygen species in natural waters. In: Foote CS, Valentine JS, Greenberg A, Liebman JF (eds) Active oxygen in chemistry. Chapman and Hall, New York, p 280
16. Mopper K, Zhou X, Kieber RJ, Kieber DJ, Sikorski RJ, Jones RD (1991) Nature 353:60
17. Benner R, Pakulski JD, McCarthyM, Hedges JI, Hatcher PG (1992) Science 255:1561
18. Hedges JI (1992) Mar Chem 39:67
19. Zepp RG (1997) Interactions of marine biogeochemical cycles and the photodegradation of dissolved organic carbon and dissolved organic nitrogen. In: Gianguzza A, Pelizzetti E, Sammartano S (eds) Marine chemistry. Kluwer Academic Publishers, Netherlands, p 329
20. Benner R (1998) Cycling of dissolved organic matter in the ocean. In: Hessen DO, Tranvik LJ (eds) Aquatic humic substances: ecology and biogeochemistry. Ecological studies, vol 133. Springer, Berlin Heidelberg New York, chap 12
21. McKnight DM, Aiken GR (1998) Sources and age of aquatic humus. In: Hessen DO, Tranvik LJ (eds) Aquatic humic substances: ecology and biogeochemistry. Ecological studies, vol 133. Springer, Berlin Heidelberg New York, chap 1
22. Perdue EM (1998) Chemical composition, structure and metal binding properties. In: Hessen DO, Tranvik LJ (eds) Aquatic humic substances: ecology and biogeochemistry. Ecological studies, vol 133. Springer, Berlin Heidelberg New York, chap 2
23. Miller WL, Moran MA (1997) Limnol Oceanogr 42(B):1317
24. Tranvik LJ (1998) Degradation of dissolved organic matter in humic waters by bacteria. In: Hessen DO, Tranvik LJ (eds) Aquatic humic substances: ecology and biogeochemistry. Ecological studies, vol 133. Springer, Berlin Heidelberg New York, chap 10
25. Blough N (1997) Photochemistry in the sea-surface microlayer. In: Liss PS, Duce RA (eds) The sea surface and global change. Cambridge University Press, England, chap 13
26. Blanchard DC, Woodcock AH (1957) Tellus 9:145
27. Blanchard DC (1978) Pure App Geophys 116:302
28. Gill PS, Graedel TE, Weschler CJ (1983) Rev Geophys 21:903
29. Cloke J, McKay WA, Liss PS (1991) Mar Chem 34:77
30. Leifer A (1988) The kinetics of environmental aquatic photochemistry. American Chemical Society Professional Reference Book, Washington DC
31. Sadiq M (1990) Toxic metal chemistry in marine environments. Marcel Dekker, New York
32. Faust BC (1994) A review of the photochemical redox reactions of iron(III) species in atmospheric, oceanic and surface waters: influences on biogeochemical cycles and oxidant formation. In: Helz GR, Zepp RG, Crosby DG (eds) Aquatic and surface photochemistry. Lewis, Boca Raton
33. Sulzberger B (1997) Effects of light on the biological availability of trace metals. In: Gianguzza A, Pelizzetti E, Sammartano S (eds) Marine chemistry. Kluwer Academic Publishers, Netherlands, p 353
34. Craig PJ, Miller D (1997) Metal ions and organometallic compounds in seawater and in sediments: biogeochemical cycles. In: Gianguzza A, Pelizzetti E, Sammartano S (eds) Marine chemistry. Kluwer Academic Publishers, Netherlands, p 85
35. Bloom PR, Leenheer JA (1989) Vibrational, electronic and high-energy spectroscopic methods for characterizing humic substances. In: Hayes MHB, MacCarthy P, Malcolm RL, Swift RS (eds) Humic substances II: in search of structure. Wiley, New York, p 411
36. Seinfeld JH, Pandis SN (1998) Atmospheric chemistry and physics: from air pollution to climate change. Wiley, New York, p26
37. Coble PG (1996) Mar Chem 51:325
38. Kouassi AM, Zika RG (1990) Neth J Sea Res 27:25
39. Green SA (1992) PhD thesis, MIT/WHOI Joint Program Oceanography
40. De Souzza Sierra MM, Donard O, Lamolte M, Belin C, Ewald M (1994) Mar Chem 47:127
41. Cauvet G, Godel F, de Souza Sierra MM, Donard O, Ewald M (1990) Cont Shelf Res 10:1025
42. Senesi N, Miano TM, Provenzano MR, Brunetti G (1991) Soil Sci 152:259
43. Miano TM, Senesi M (1992) Sci Total Environ 117/118:41

44. Peuravuori J, Pihlaja K (1997) Anal Chim Acta 337:133
45. Klinkhammer GP, Chen CS, Wilson C, Rudnicki MD, German CR (1997) Mar Chem 56:1
46. Green SA, Blough NV (1994) Limnol Oceanogr 39:1903
47. Ferrarri G, Mingazzini M (1995) Mar Ec Prog Ser 125:305
48. Mopper K, Schultz CA (1993) Mar Chem 41:229
49. Mobed JJ, Hemmingsen SL, Autry JL, Mcgown LB (1996) Environ Sci Tech 30:3061
50. Milne PJ, Odum DS, Zika RG (1987) Time-resolved fluorescence measurements on dissolved organic matter. In: Zika RG, Cooper WJ (eds) Photochemistry of environmental aquatic systems. ACS Symposium Series 237, American Chemical Society, Washington DC, chap 10
51. Milne PJ, Zika RG (1989) Luminescence quenching of dissolved organic matter in seawater. Mar Chem 27:147–164
52. Lochmuller CH, Saavedra SS (1986) Anal Chem 58:1978
53. Power JF, LeSage R, Sharma DK, Langford CH (1986) Env Tech Lett 7:425
54. Cook RL, Langford CH (1995) Anal Chem 67:174
55. Chen S, Inskeep WP, Williams SA, Callis PR (1994) Environ Sci Tech 28:1582
56. Jones G II, Indig GL (1996) New J Chem 20:221
57. Hemmingsen SL, Mcgown LB (1997) Appl Spec 51:921
58. Filippova EM, Boichuk IV, Dolenko TA, Fadeev VV (1997) Proc 3rd EARSeL Workshop on Lidar Remote Sensing of Land and Sea, p 51
59. Minero C (1997) Light and chemically driven reactions and equilibria in the presence of organic and inorganic colloids. In: Gianguzza A, Pelizzetti E, Sammartano S (eds) Marine chemistry. Kluwer Academic Publishers, Netherlands, p 39
60. Sharp JH, Benner R, Bennet L, Carlson CA, Fitzwater SE, Peltzer ET, Tupas LM (1995) Mar Chem 48:91
61. Wotton RS (1990) The biology of particles in aquatic systems. Lewis, Boca Raton
62. Hurd DC, Spencer DW (eds) (1991) Marine particles: analysis and characterization. Geophys Monograph 63, American Geophysical Union, Washington DC
63. Petronio BM (1997) Techniques of extraction and analytical methods for humic substances in seawater and sediments. In: Gianguzza A, Pelizzetti E, Sammartano S (eds) Marine chemistry. Kluwer Academic Publishers, Netherlands, p 211
64. Choudry GG (1983) Humic substances: structural, photophysical, photochemical and free radical aspects and interactions with environmental chemicals. Gordon and Breach, New York
65. Hayes MHB, MacCarthy P, Malcolm RL, Swift RG (1989) Structures of humic substances: the emergence of "forms". In: Hayes MHB, MacCarthy P, Malcolm RL, Swift RS (eds) Humic substances II: in search of structure. Wiley, New York, p 690
66. Amador JA, Milne PJ, Moore CA, Zika RG (1990) Mar Chem 27:147
67. Benner R (1991) Ultrafiltration for the concentration of bacteria, viruses and dissolved organic matter. In: Hurd DC, Spencer DW (eds) Marine particles: analysis and characterization. Geophysics. Monograph 63, American Geophysical Union, Washington DC, p 181
68. Chin Y-P, Aiken G, O'Loughlin E (1994) Env Sci Tech 28:1853
69. Zanardi E, Moore C, Clark CD, Jimenez J, Jones G, Zika RG (1999) Fluorescence lifetimes of size-fractionated CDOM in a South Florida riverine to marine transition zone. Special session: Composition and Reactivity of DOC: Comparisons Across Freshwater and Marine Environments. Annual Meeting of the American Society of Limnology and Oceanography, Santa Fe, USA
70. Martin M, Williams PS (1992) Theoretical basis of field-flow fractionation. In: Dondi F, Guichon G (eds) Theoretical advancement in chromatography and related separation techniques. Kluwer Academic Publishing, Netherlands, p 513
71. Beckett R, Jue Z, Giddings JC (1987) Environ Sci Tech 21:289
72. Beckett R, Bigelow JC, Jue Z, Giddings JC (1989) Analysis of humic substances using flow field flow fractionation. In: Suffet JH, MacCarthy P (eds) Aquatic humic substance: influences on fate and treatment of pollutants. ACS Symposium Series 219, American Chemical Society, Washington DC, p 65

73. McCarthy MD, Hedges JI, Benner R (1993) Chem Geol 107:503
74. Benner R, Biddanda B, Black B, McCarthy M (1997) Mar Chem 57:243
75. Pakulski JD, Benner R (1994) Limnol Oceanogr 39:930
76. Vaughn PP, Blough NV (1998) Environ Sci Technol 32:2947
77. Vincenti M (1997) Application of mass spectrometric techniques to the detection of natural and anthropogenic substances in the sea. In: Gianguzza A, Pelizzetti E, Sammartano S (eds) Marine chemistry. Kluwer Academic Publishers, Netherlands, p 189
78. MacCarthy P, DeLuca SJ, Voorhees KJ, Malcolm RL, Thurman EM (1985) Geochim Cosmochim Acta 49:2091
79. Saiz-Jiminez C (1994) Environ Sci Tech 28:1773
80. Zepp RG (1988) Environmental photoprocesses involving natural organic matter. In: Frimmel FH, Christman RF (eds) Humic substances and their role in the environment. Wiley, New York, p 1983
81. Pos WH, Milne PJ, Riemer DD, Zika RG (1997) J Geophys Res 102:12, 831
82. Pos WH, Milne PJ, Riemer DD, Zika RG (1998) Mar Chem (submitted)
83. Pos WH, Riemer DD, Zika RG (1998) Mar Chem 62:89
84. Power JF, Sharma DK, Langford CH, Bonneau R, Jossert-Dubien J (1987) Laser flash photolysis studies of a well-characterized soil humic substance. In: Zika RG, Cooper WJ (eds) Photochemistry of environmental aquatic systems. ACS Symposium Series 237, American Chemical Society, Washington DC, p 157
85. Zepp RG, Braun AM, Hoigne J, Leenheer JA (1987) Environ Sci Tech 21:485
86. Swallow AJ (1969) Nature 222:369
87. Silva C, Walhout PK, Yokoyama K, Barbara PF (1998) Phys Rev Letts 80:1086
88. Lean DRS, Cooper WJ, Pick FR (1994) Hydrogen peroxide formation and decay in lake waters. In: Helz GR, Zepp RG, Crosby DG (eds) Aquatic and surface photochemistry. Lewis, Boca Raton, chap 16
89. Micinski E, Ball LA, Zafiriou OC (1993) JGR 98:2299
90. Petasne RG, Zika RG (1987) Nature 325:516
91. Petasne RG, Zika RG (1997) Mar Chem 56:215
92. Moffett JW, Zafiriou OC (1993) JGR 98:2307
93. Moore CA, Farmer CT, Zika RG (1993) J Geophys Res 98:2289
94. Sikorski RJ, Zika RG (1993) J Geophys Res 98:2315
95. Sikorski RJ, Zika RG (1993) J Geophys Res 98:2329
96. Scully NM, McQueen DJ, Lean DRS (1996) Limnol Oceanogr 41:540
97. Fujiwara K, Ushiroda T, Takeda K, Kumamoto YI, Tsubota H (1993) Geochem J 27:103
98. Burns SE, Hassett JP, Rossi MV (1996) Environ Sci Technol 30:2934
99. Burns SE, Hassett JP, Rossi MV (1997) Environ Sci Technol 31:1365
100. Voelker BM, Morel FMM, Sulzberger B (1997) Environ Sci Tech 31:1004
101. Weber JH (1988) Binding and transport of metals by humic materials. In: Frimmel FH, Christman RF (eds) Humic materials and their role in the environment. Wiley-Interscience, New York, p 165
102. Suffet JH, MacCarthy P (eds) (1989) Aquatic humic substances: influence on fate and treatment of pollutants. Advances in Chemistry Series, 219, American Chemical Society, Washington DC
103. Zafiriou OC, True MB (1979) Mar Chem 8:9
104. Zafiriou OC, Bonneau R (1987) Photochem. Photobiol 45:723
105. Daniels M, Meyers RV, Belardo EV (1968) Symp Inorg Photochem 72:389
106. Zhou X, Mopper K (1990) Mar Chem 30:71
107. Mopper K, Zhou X (1990) Science 256:661
108. Anastasio C, Allen JM, Faust B (1994) Aqueous-phase photochemical formation of peroxides in authentic cloud waters. In: Helz GR, Zepp RG, Crosby DG (eds) Aquatic and surface photochemistry. Lewis, Boca Raton, chap 18
109. Faust BC, Allen JM (1992) J Geophys Res 97:12, 913
110. Faust BC, Anastasio C, Allen JM, Arakaki T (1993) Science 260:73
111. Faust BC, Hoigne J (1990) Atmos Env 24 A:79

112. Carlson DJ (1982) Nature 286:482
113. Carlson DJ (1983) Limnol Oceanogr 28:415
114. O'Dowd CD, Smith MH, Consterdine IE, Lowe JA (1997) Atmos Env 31:73
115. Murphy DM, Anderson JR, Quinn PK, McInnes LM, Brechtel FJ, Kreidenweiss SM, Middlebrook AM, Posfai M, Thomson DS, Buseck PR (1998) Nature 392:62
116. Lean D (1998) Attenuation of solar radiation in humic waters. In: Hessen DO, Tranvik LJ (eds) Aquatic humic substances. Ecological studies vol. 133, Springer, Berlin Heidelberg New York, chap 5
117. Seinfeld JH, Pandis SN (1998) Atmospheric chemistry and physics: from air pollution to climate change. Wiley, New York, p 441
118 Seinfeld JH, Pandis SN (1998) Organic Aerosols. In: Atmospheric chemistry and physics: from air pollution to climate change. Wiley, New York, chap 13
119. Mopper K, Zika RG (1987) Nature 325:246
120. Milne PJ, Zika RG (1993) J Atmos Chem 16:361
121. Willey JD, Kieber RJ, Eyman MS, Avery GB Jr (1999) Global biogeochem cycles (in press)
122. Zafiriou O, Gagosian RB, Peltzer ET, Alford JB, Loder TT (1985) J Geophys Res 90:2409
123. Sempere R, Kawamura K (1996) Atmos Environ 30:1609
124. McDowell WH, Sanchez CG, Asbury CE, Perez RR (1990) Atmos Environ 24 A:2813
125. Ekllund TJ, McDowell WH, Pringle CM (1997) Atmos Environ 31:3903
126. Zhou X, Mopper K (1997) Mar Chem 56:201
127. Zika RG, Saltzman ES, Chameides WL, Davis DD (1982) J Geophys Res 87:5015
128. Cooper WJ, Saltzman ES, Zika RG (1987) J Geophys Res 92:2970
129. Saltzman ES (1986) PhD thesis, Univ of Miami
130. Allen JM, Faust B (1994) Aqueous-phase photochemical formation of peroxyl radicals and singlet molecular oxygen in cloud water samples from across the United States. In: Helz GR, Zepp RG, Crosby DG (eds) Aquatic and surface photochemistry. Lewis, Boca Raton, chap 18
131. Graedel TE, Mandich ML, Weschler CJ (1986) J Geophys Res 91:5205
132. Zuo Y (1995) Geochim Cosmo Acta 59:3123
133. Harvey RG (1998) Environmental chemistry of PAHs. In: Neilson AH (ed) The handbook of environmental chemistry, vol 3, part 1: PAHs and related compounds. Springer, Berlin Heidelberg New York, chap 1
134. Berlaman IB (1971) Handbook of fluorescence spectra of aromatic molecules. Academic Press, New York
135. Arey J (1998) Atmospheric reactions of PAHs including formation of nitroarenes. In: Neilson AH (ed) The handbook of environmental chemistry, vol 3, part 1: PAHs and related compounds. Springer, Berlin Heidelberg New York, chap 9
136. Dalsey JM, Lewandowski CG, Zorz M (1982) Environ Sci Technol 16:857
137. Jang M, Mcdow S (1995) Environ Sci Technol 29:2654
138. Kamens RM, Guo Z, Fulcher JN, Bell DA (1988) Environ Sci Technol 22:103
139. Rudnick SM, Chen RF (1998) Talanta 47:907

CHAPTER 2

Colloids and the Ocean Carbon Cycle

Paul E. Kepkay

Biological Oceanography Section, Ocean Sciences Division, Fisheries and Oceans Canada, Bedford Institute of Oceanography, P.O. Box 1006, Dartmouth, Nova Scotia B2Y 4A2, Canada
E-mail: KepkayP@mar.dfo-mpo.gc.ca

Colloidal organic carbon (COC) is a globally significant fraction of dissolved organic carbon (DOC) in the surface ocean. As a non-living but reactive reservoir of carbon, COC outweighs the living carbon stored in marine biomass by a wide margin. Aggregation, respiration and photooxidation are all involved in the removal of COC from surface waters. Despite these removal mechanisms, bioreactive COC can still build up. The end result of the incorporation of this COC into the repeated diurnal cycling of carbon is the accumulation of old, low molecular weight organic carbon (LMWOC) at the expense of younger, more reactive COC. With respect to the sequestration of carbon in deep water, COC is either caught up into aggregates that can sediment out to fuel the deep ocean ecosystem or is converted to less reactive LMWOC that can accumulate prior to deep transport by winter mixing. An understanding of the degree to which aggregation or accumulation is coupled to the net production of DOC is a key requirement for any quantitative description of the ocean carbon cycle.

Keywords. Colloidal organic carbon, Dissolved organic carbon, Bioreactivity, Photoreactivity.

The Handbook of Environmental Chemistry Vol. 5 Part D
Marine Chemistry
(ed. by P. Wangersky)

1 Introduction

The ocean is one of the largest reservoirs of organic (biogenic) carbon on the earth's surface [1]. Globally, the composition of this reservoir is regulated by feedback mechanisms that maintain the planetary environment within the relatively narrow range necessary for life [2, 3]. Phytoplankton productivity is the ultimate source of organic carbon in the ocean [4]: the best estimate of this primary production is 45–50 Gt C year^{-1} [5] – about half of the combined total for terrestrial and marine productivity [1].

In contrast to terrestrial systems [1], very little of the carbon produced by the phytoplankton remains as living carbon. For example, when primary production is considered in relation to the rate at which it is turned over by decomposition (Table 1), the standing stock of phytoplankton is small (1–11 Gt C). The majority (>90%) of the organic carbon accumulates as two non-living fractions: detritus and dissolved organic carbon (Fig. 1). Dissolved organic carbon is the predominant non-living carbon pool (Fig. 1), and the majority of this carbon resides in a low molecular weight size fraction (Fig. 2) that appears remarkably resistant to biological decomposition [4]. Even though the high molecular weight or colloidal size fraction is less than half (10–40%) of the dis-

Table 1. Standing stock of phytoplankton biomass in the world's ocean compared with primary production and decomposition of the primary production by respiration

Primary (phytoplankton) production[a] (Gt C year^{-1})	Carbon turnover by respiration (days)	Standing stock of phytoplankton[b] (Gt C)
45	10[c]	1.2
50	80[d]	11.0

[a] Upper and lower limits of primary production from Longhurst et al. [5].
[b] Calculated as the product of primary production and carbon turnover.
[c] Lower limit of (fastest) carbon turnover from Eppley et al. [136].
[d] Upper limit of (slowest) carbon turnover from Benner [4].

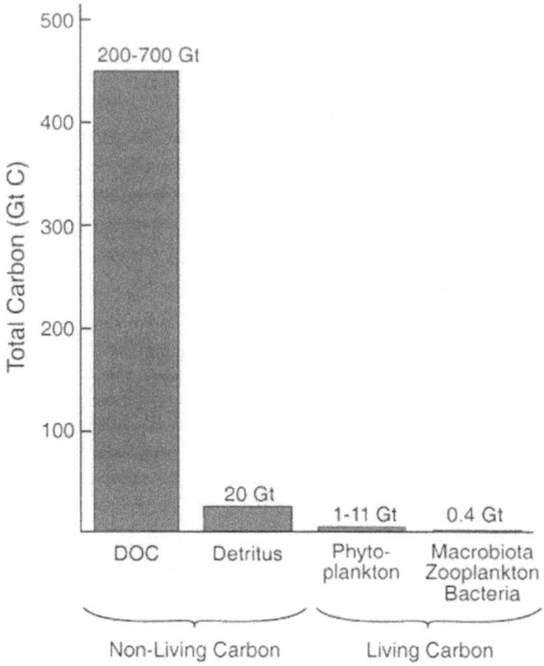

Fig. 1. Dissolved organic carbon (DOC) and the other major reservoirs of living and non-living organic carbon in the world's ocean. DOC, zooplankton and bacterial biomass are from Cauwet [137] and Siegenthaler and Sarmiento [138]; phytoplankton biomass is from Table 1; macrobiota biomass is from Stewart [139]

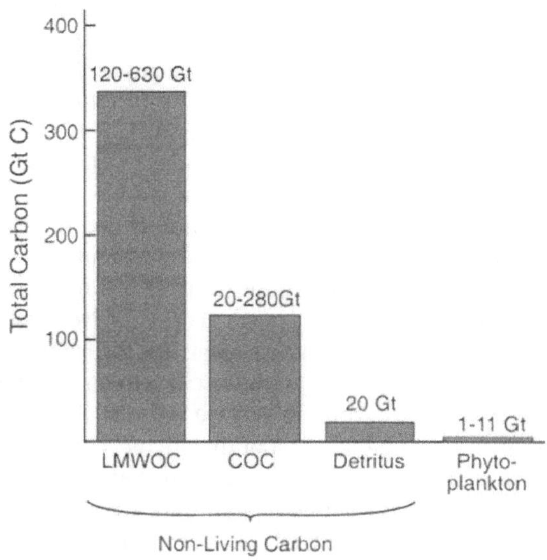

Fig. 2. Subdivision of dissolved organic carbon (DOC) into two submicron size fractions – colloidal organic carbon (COC) and low molecular weight organic carbon – using the results from ultrafiltration studies [9, 12–21]

solved organic carbon [4, 6], this colloidal organic carbon is a good deal more bioreactive than the low molecular weight material [6–9].

Not only is the colloidal fraction the most bioreactive component of dissolved organic carbon, it is also inextricably linked by colloid aggregation to the cycling of particulate size fractions [6]. If colloids, their reactivity and their ability to aggregate are not taken into account, an incomplete picture of the ocean carbon cycle is the result. In this chapter, a more complete picture will be developed by focusing on the multifaceted contribution of colloids to the accumulation and deep transport of organic carbon.

2 Size Fractionation – Definition of DOC and COC

The major components of organic carbon in the ocean are dissolved and particulate organic carbon (DOC and POC, respectively), amounting to more than 80% of the total. The remainder is deposited in surface sediments on river deltas and continental shelves [1]. POC, which is made up predominantly of detritus (Fig. 1), is defined solely on the basis of size. It is generally measured as the carbon retained on filters with nominal pore sizes of 0.5–0.7 μm and accounts for <5% of the organic carbon in seawater [4]. Even though most of the organic carbon is in the so-called dissolved fraction that passes through a filter, it is important to keep in mind that DOC is also defined on the basis of size rather than solubility or other physicochemical characteristics. In addition, the actual concentrations of DOC are low – in the region of 80–100 μM C in surface water and 30–50 μM C in deep water [10, 11].

When seawater is further size fractionated by cross (or tangential) flow ultrafiltration [9, 12–21], 10–40% of the DOC (Fig. 2) is retained by membranes with molecular weight cutoffs of 1–10 kDa (equivalent to spherical diameters of 1–3 nm), and is referred to as high molecular weight or colloidal organic carbon (COC). Most of the DOC passes through ultrafiltration membranes to be measured as low molecular weight organic carbon (LMWOC). Within the colloidal size range of 1–10 kDa at the lower end and 0.2–1.0 μm at the upper end, the size distributions of organic matter follow Pareto laws, with slopes of between −2 and −3 [22, 23]. These size distributions are not representative of individual particles. Instead they are combined distributions of colloids and their aggregates [24] that are far from uniform in shape and size.

To obtain an idea of how the four (particulate, dissolved, colloidal and low molecular weight) size classes of organic carbon are distributed, idealized profiles of their concentration over depth in the Pacific Ocean are outlined in Fig. 3. The concentration of organic carbon in surface water is about double the deep water concentration, indicating that most of the organic carbon originates in the surface ocean. As depth increases, size distribution changes; the larger size classes (POC and COC) account for about 30% of the carbon in surface water and decrease to about 20% below a depth of 200 m. Even though higher concentrations of POC and COC are restricted to depths shallower than 200 m (Fig. 3), most of the organic carbon in the ocean resides in deep water [4]. Nearly all of this carbon is DOC and about 80% of the deep water DOC is LMWOC.

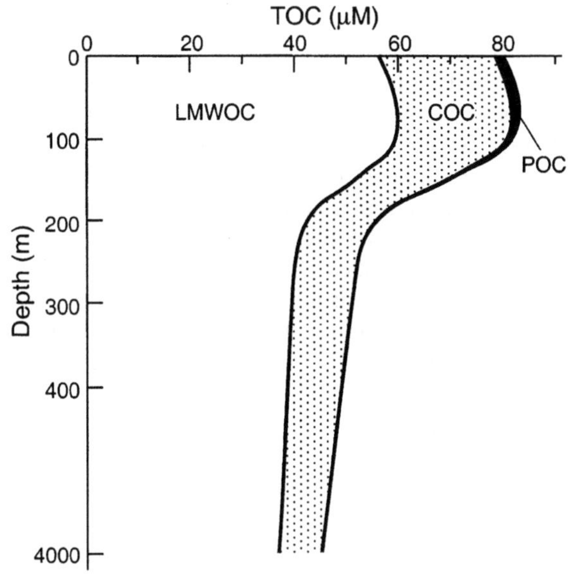

Fig. 3. Depth distribution of three size classes of organic carbon at 12 °S, 135 °W in the Pacific Ocean. Low molecular weight organic carbon (LMWOC) is < 1–3 nm (1–10 kDa) in size; colloidal organic carbon (COC) is between 1–3 nm and 700 nm (0.7 μm); and particulate organic carbon (POC) is > 700 nm in size. Distributions based on the data of Benner et al. [16]

3 Origin and Age of DOC and COC

A considerable amount of analytical effort has been expended to determine the origin and age of organic matter in seawater. Most of the work has been focused on the radioisotope content of either bulk DOC or of the colloidal fractions isolated by ultrafiltration. This means that the origin or age of the LMWOC that passes through an ultrafiltration membrane must be interpolated from the "average" or total isotopic signature of DOC by accounting for the signature of COC.

3.1 Origin

Most of the DOC in the ocean appears to be of marine origin, with terrestrial organic carbon accounting for only 1% of the total annual production [4]. Stable carbon isotope ($\delta^{13}C$) ratios of marine DOC are in the range –20 to –33 ppt [25, 26], very similar to the ratios measured for marine phytoplankton [27] and enriched in ^{13}C compared to the ratios of –27 to –29 ppt recorded for riverine carbon [28]. In addition, measurements of lignin-derived phenols (unique biomarkers of terrestrial plant carbon) indicate that terrestrial carbon is < 5% of the DOC in the ocean [29, 30].

Phytoplankton may be the ultimate source of marine DOC, but the individual importance of each of the many and diverse mechanisms of DOC formation is not known. A comprehensive review of these mechanisms will not be attempted here, but it is still important to acknowledge that the composition of both DOC and COC will be governed by how they are generated. Potentially important mechanisms include direct exudation by phytoplankton [31], cell lysis during autolysis [32] or viral infection [33], and release from cells in response to grazing [34]. Both DOC and COC can also be released directly as extracellular exudates by the phytoplankton [18, 19]. The substantial carbohydrate (polysaccharide) content of these exudates [35] may allow them to be a source of bioreactive organic carbon that can be decomposed rapidly by heterotrophic activity [4, 15]. Grazing and cell lysis may also generate bioreactive organic carbon by releasing the internal components of cells as proteins [36–38].

3.2 ^{14}C Age

Even though most of the organic matter that is produced in the ocean is recycled rapidly (Table 1), the ^{14}C age of DOC in the deep ocean is 4000–6000 years BP [39, 40]. One of the few concrete conclusions that can be drawn from this apparent contradiction is that marine DOC is a highly diverse pool of organic molecules that is turned over at a variety of rates [4]. This is confirmed in the case of colloidal fractions isolated from surface waters [41]; COC in the size range of 10 kDa to 0.2 μm is younger than 40 years BP, whereas COC in the range of 1 kDa to 0.2 μm is between 380 and 4500 years BP. Taken as a whole, the ^{14}C ages suggest that COC is a mixture of several end members, with large colloids turned over rapidly and smaller colloids turned over slowly. In addition, the data converge on LMWOC as the oldest fraction of DOC, with turnover times on the order of thousands of years.

4 Composition of DOC and COC

Over the last 50 years, analysis of the individual constituents of DOC in seawater has proven to be challenging to say the least. Recent progress in the field has been based on three approaches: the elemental analysis of dissolved organic material, the isolation and spectroscopic analysis of specific fractions of DOC, and the direct analysis and electron microscopy of specific biochemical components of DOC.

4.1 Carbon to Nitrogen Ratio

A wide range of C/N ratios have been reported for marine dissolved organic matter [9, 42–46]. Most values are between 10 and 30, indicating that dissolved organic matter is depleted in nitrogen compared to the ratios of between 6 and 10 reported for particulate material [47]. Even though increased C/N ratios can be attributed to the production of colloidal exudates by phytoplankton during blooms [19], it is not clear if the depletion in nitrogen is related solely to the presence of carbon-rich components such as carbohydrates [15, 35] or also includes the selective and preferential utilization of nitrogenous components during decomposition.

4.2
Carbohydrates and Humics – Nuclear Magnetic Resonance (NMR) Spectroscopy

A variety of spectroscopic techniques have been applied to DOC isolated from seawater by cross-flow ultrafiltration or adsorption onto XAD resins. The two techniques isolate very different organic fractions from seawater. Hydrophobic fractions (such as marine humic material) are isolated on XAD resins [48], whereas the organic matter extracted by ultrafiltration is retained primarily on the basis of its molecular size and shape [49], resulting in isolates rich in nitrogen and carbohydrates (polysaccharides). Nuclear magnetic resonance (NMR) spectroscopy has proven successful in distinguishing between the specific structures of XAD-bound humics and the carbohydrates concentrated into colloidal size fractions.

The solid-state ^{13}C NMR spectra of marine humics are characterized by unsubstituted alkyl carbons [50–52], whereas freshwater humics have a much higher aromatic carbon content [51–53], reflecting the input of material derived from vascular plants. Given that XAD resins recover only 5–25% of the DOC in seawater [26, 54], the marine humics characterized by NMR spectroscopy are probably not a major component of DOC. In addition, the similarities in concentration and structural characteristics of humics from a variety of marine environments suggest that it is relatively unreactive. This overall lack of reactivity is confirmed by the ^{14}C ages of marine humics, which are older than the average (total) ^{14}C age of DOC [26].

The structural features of the COC isolated by ultrafiltration are very different from the structure of marine humics. Solid-state ^{13}C NMR spectra of COC reveal a predominance of oxygen-substituted alkyl carbons that are characteristic of carbohydrates [15]. The NMR spectra also indicate that 11 μM C (about half) of the COC in surface water is associated with polysaccharides [4]; in deep water, only 2.2 μM C is COC associated with polysaccharides. A five-fold decrease in carbon associated with carbohydrates is a strong indication that a major fraction of DOC is cycled in the upper ocean [15]. At this point, however, it should be stressed that ultrafiltration recovers between 10 and 40% of the carbon in DOC [4, 6]. There is still a large fraction (60–90%) of marine DOC that cannot be identified, although Benner [4] has speculated that between 30 and 50% of DOC could be recovered from seawater using a combination of ultrafiltration and XAD extraction.

4.3
Carbohydrates and Amino Acids – Direct Analysis and Electron Microscopy

Direct analyses of the individual components of marine DOC have been difficult due to their low concentrations in association with high concentrations of salts. In spite of these problems, sensitive methods have been developed for the direct analysis of two biotic components – amino acids and carbohydrates (polysaccharides). Dissolved free and combined amino acids in seawater range from 300–500 nM in surface water and from 100–150 nM in the deep ocean [26, 55]. These concentrations are about 2% of DOC in surface water and 1% of deep water DOC.

Through application of a colorimetric method [56] in combination with an acid hydrolysis procedure [57], carbohydrates have been found to be in the range of 2–5 µM glucose equivalents in surface waters and 0.8–1.1 µM in the deep ocean [58]. When expressed as organic carbon [4, 59], the carbohydrates amount to the largest single identifiable fraction of organic carbon in seawater, accounting for about 25% of the DOC in surface waters and <10% of DOC in the deep ocean. The compositional and structural importance of fibrillar polysaccharides in the colloidal size fraction are further illustrated in electron micrographs of the colloidal material collected by ultrafiltration [60]. As a visual record, these micrographs highlight the structural complexity of the colloidal size fraction, with compact inorganic and organic colloids suspended in an interlinked network of polysaccharides and other biopolymer fibrils [24].

Given that colloidal carbohydrates in seawater may exist as interlinked fibrils of polysaccharide, their true molecular weight will be difficult to define by ultrafiltration. In addition, marine carbohydrates include a diverse group of monomers and polymers, most of which have not been identified [4]. Aldoses, however, are an exception and have been quantitatively measured in seawater [61, 62], ranging in concentration from 300–800 nM in surface waters and from 120–170 nM in the deep ocean [63, 64]. Even though these concentrations are only 10–20% of the total carbohydrate pool and <3.5% of DOC, Benner [4] has pointed out that, by virtue of their distribution in the various size fractions of DOC, the aldoses may be a sensitive indicator of the degree of diagenetic alteration or "freshness" of organic carbon in the ocean.

5 Aggregation of COC

The importance of colloids in the ocean lies not only in their role as a global pool of organic carbon, but also in their ability to act as sites for the adsorption of trace metals and organic compounds [24]. Between 40 and 90% of trace compounds can be scavenged by adsorption onto marine colloids [65]. If the colloids remain in suspension, both their organic carbon and associated trace compounds can be transported and redistributed over long distances and times. Nevertheless, the range of structures and reactive sites in natural colloids leads almost inevitably to their aggregation and the formation of small (submicron) aggregates [24]. In turn, these small aggregates can coagulate to produce larger aggregates that are hundreds of µm to mm across and are capable of rapid sedimentation.

5.1
Homo- vs Heteroaggregation

Aggregation is generally thought to be a two-step process; particles must first collide and then stick together. As a result, a great deal of attention has been paid to the role of collision efficiency and particle stickiness [6]. In addition, specific collision mechanisms, such as Brownian motion, turbulent shear and surface coagulation, have been examined in some detail [6, 66]. However, as Buffle et al. [24] have pointed out, a general physicochemical theory exists only for the homoaggregation of compact (near-spherical) colloidal particles that aggregate with each other. Based on the Smolouchowski equations for Brownian motion [67] and DLVO theory combined with short-range hydrodynamic forces [68–71], this approach has been useful for predicting the behavior of compact particles with similar physicochemical properties and diameters >100 nm [72, 73]. Even though there is a growing amount of literature on the properties of colloidal polymers and their interactions in natural waters [74, 75], there is no similar theory for the heteroaggregation of different particles, especially for the aggregation of polymer chains and compact colloids [24].

5.2
Colloids as a Three-Component System

By reviewing microscopic observations, physicochemical properties and numerical simulations of colloid dynamics, Buffle et al. [24] have developed a three-component picture of the heteroaggregation of colloids. In marine systems the three components are inorganic colloids (IC), refractory organic matter (ROM) with characteristics similar to freshwater fulvic material, and fibrillar polymers (FP).

Whether or not these components interact to produce aggregates that are large enough to sediment out of the water column depends on their relative amounts and physicochemical states. As a general approximation, Buffle et al. [24] suggest that IC–ROM and FP–ROM interactions result in small (submicron) aggregates that remain in suspension (Fig. 4). Weeks or months are required for aggregates of IC or ROM to reach sizes that are large enough to induce sedimentation. In addition, ROM coatings on IC and FP impose an overall negative charge on the larger colloids, resulting in slow reaction-limited aggregation. In contrast, IC–FP interactions generate aggregates by polymer bridging; the aggregates are either large enough (on the order of 100 μm–mm across) to directly sediment out of the water column (Fig. 4) or are capable of coagulating into larger aggregates in response to the input of additional kinetic energy by turbulent shear or surface coagulation [6, 66]. Added to this is the fact that, during blooms, the degree of aggregation is related to the amount of FP released by the phytoplankton as sticky polysaccharides [76]. This means that aggregation rates will not only vary seasonally, but will also be regulated by two opposing processes: the stabilization of small aggregates by slow aggregation rates and ROM coatings, and the generation of large sedimenting aggregates by

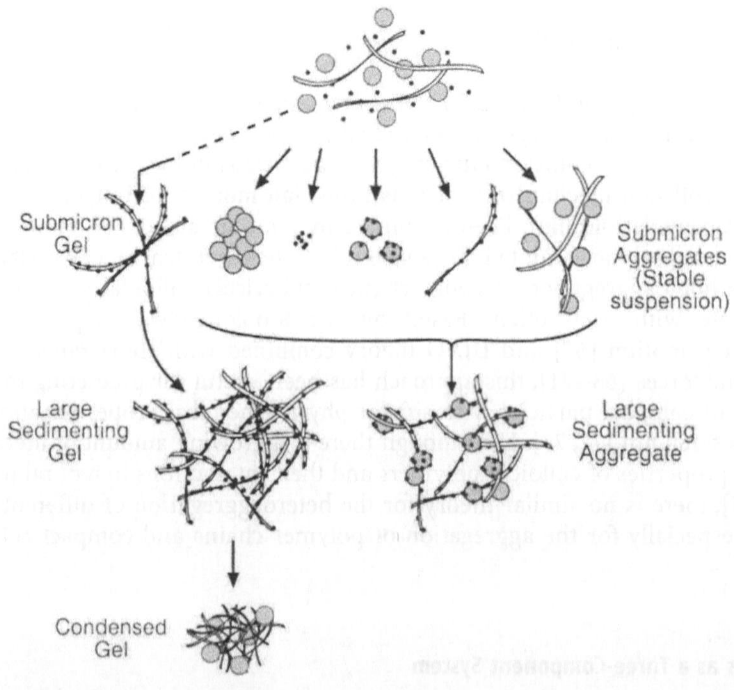

Fig. 4. Aggregates formed in a three-component, colloidal system [24]. Refractory (fulvic-like) organic matter (ROM) is represented by *small dots*, inorganic colloids (IC) by *larger, shaded spheres* and fibrillar polymers (FP) by *ribbons*. In addition to small (submicron) and larger (>1 mm) aggregates, refractory organic matter and fibrillar polysaccharides can also form gels [78] where inorganic colloids can develop internally after condensation

the bridging of IC by FP. In summary, Buffle et al. [24] have suggested that the likelihood of producing small and stable aggregates (that remain in suspension) as opposed to large aggregates (that sediment out of the water column) is controlled by the ratio of IC to FP and the degree of surface coverage by ROM.

5.3
Polymer Gel Formation and Condensation

Gels are a three-dimensional network of polymers and a solvent which, in the case of hydrogels, is water [77]. The water within a hydrogel supports a network of fibrils that traps the water and creates a microenvironment in thermodynamic equilibrium with the surrounding medium [78]. The kinetic energy required for the so-called "spontaneous" formation of gels from colloids [78] has its origins in Brownian motion; in natural seawater, 5 μm gels can be generated from DOC in 50–83 h [78]. With additional seawater and in the absence of disaggregation, microgels (in the submicron size range) will continue to form by the assembly of free polymers and go on to produce macrogels by aggregation and annealing processes (Fig. 4). Chin et al. [78] have also found that marine gels exhibit reversible phase transitions – a behavior that is typical of gels in general. Microgels swell as seawater pH is raised and condense as pH is lowered. In addition, the fact that gel condensation takes place at pH 4.5 (near the dissociation constant of carboxylic acid) is consistent with observations that about 25% of DOC is carboxyl groups associated with polysaccharide structures [4, 15].

Given that the polymer matrix of marine gels is a tangled network of fibrils stabilized by cation bonds rather than a structured cross-linked network [78], there seems to be little functional difference between these gels (Fig. 4) and the types of aggregate proposed by Buffle et al. [24]. However, the discovery by Chin et al. [78] of bound Ca^{2+} within marine gels leads to the possibility that inorganic, calcium-rich colloids can be generated internally by small increases in pH and/or gel condensation. This leads to the possibility that marine gels can generate their own inorganic "ballast", resulting in rapidly sedimenting aggregates (Fig. 4). When combined with the fact that the polymer matrix of the microgels contains proteins and lipids as well as carbohydrates [78], these aggregates may also be a rich source of bioreactive organic matter.

5.4 Transparent Exopolymer Particles

Even though the selective application of stains to highlight specific organic aggregates has been in use for some time [79], Alldredge et al. [80] have utilized Alcian blue in a new way to define a class of transparent exopolymer particles (TEPs) that is generated abiotically. TEPs are, in fact, the aggregates formed by coagulation of colloidal fractions of DOC in response to the input of kinetic energy by fluid shear [80, 81] or bubbling [6, 62]. The high carbohydrate content of TEPs [62] suggests that they may be produced by coagulation of the small (submicron) aggregates of Buffle et al. [24] or the microgels of Chin et al. [78] in Fig. 4. In addition, TEPs are found in a wide range of marine environments [81–85] in relatively large numbers (5000–50,000 ml^{-1}).

The overall importance of TEPs in the ocean's carbon cycle is two-fold: (i) by acting as sticky intermediaries [86] they enhance the formation of large, rapidly sedimenting aggregates and (ii) the consumption of TEPs by protozoans [87, 88], larvaceans [89], and perhaps even by copepods [90, 91], allows the marine food web to tap into colloidal fractions of DOC – a resource that would not normally be exploited in the absence of aggregation [6, 66]. The production of the colloidal precursors of TEPs by phytoplankton during blooms is also a "short-circuiting" of decomposition pathways in the microbial loop [9] – transferring energy from DOC to the marine biota without first having to regenerate the DOC by decomposition of larger particulate material.

The formation of TEPs may transfer more than just reactive carbon to the ocean ecosystem. As Mari [85] has pointed out, the composition of TEPs can be influenced by the adsorption of amino acids [92] and metals [93]. In addition, Long and Azam [94] have identified Coomassie stained particles (CSPs) as protein-rich aggregates in seawater. This means that, even though the process of colloid aggregation can exert a strong influence on the cycling of bioreactive carbon through the production of small and large aggregates (Fig. 4), it may have equally far-reaching effects on the cycling of nitrogen and trace metals.

5.5
Disaggregation

Disaggregation is at least as important as aggregation and can be triggered by physical or biological processes (or more likely by a combination of the two). Unfortunately, disaggregation is not often incorporated into models predicting aggregate size [73] because the data are not available to accurately define the kinetics of disaggregation. Even so, Jackson [95] has demonstrated that, in order for a model to predict aggregate size in a mesocosm, the effects of disaggregation must be taken into account. With respect to size, Hill [96] has noted that aggregates smaller than the smallest eddy in turbulent flow will not experience much in the way of disaggregation. On this basis, he developed a model where aggregate erosion rate depends on particle size. Results from the model indicate: (i) the splitting of aggregates rather than their progressive erosion is the predominant disaggregation process, and (ii) aggregates $<100\ \mu m$ across do not suffer appreciable breakup. By implication these results suggest that, once large aggregates are generated by the coagulation of smaller (submicron) aggregates (Fig. 4), the colloidal components of the aggregates are moved more-or-less in one direction – up the size spectrum. However, as Jackson and Burd [73] have pointed out, the strength of the physicochemical bonds that hold an aggregate together are poorly known. They go on to caution that bacteria and zooplankton can destabilize aggregates to the point where they break apart at relatively low shear rates. More specifically, the action of bacterial ectoenzymes [97–99] and copepod feeding [90, 91] may act in combination to induce the breakup of aggregates on a variety of scales.

6
Reactivity of COC and DOC

Kepkay [6] has pointed out that the reactivity of DOC as a whole (and COC in particular) should be accounted for to obtain an accurate estimate of carbon cycling between the surface and the deep ocean. In mid and deep water, this reactivity can be defined solely in terms of the bioreactivity of organic carbon – more specifically, the "burning off" of carbon to CO_2 by respiration. In surface waters, photochemical oxidation [100, 101] acts as an additional source of CO_2. This means that the reactivity of COC must be defined in terms of both bio- and photoreactivity before the amount of carbon available for deep transport and sequestration can be determined.

6.1
Bioreactivity

6.1.1
Microbial Respiration

The microbial community (including the bacteria and protozoa) is regarded as the predominant consumer of DOC in the ocean [102], utilizing a large fraction of daily primary production. Bacterial production (the net heterotrophic production of bacterial biomass) has been used as an estimate of the productivity of the microbial community as a whole, with rates of bacterial production in surface waters in the region of 240 nM C day^{-1} [103]. In addition, most estimates [6, 104–106] have established bacterial growth efficiency at <20%. This means that the DOC respiration calculated from these measurements (1.2 μM C day^{-1}) is at least five times greater than bacterial production [4]. This estimate of DOC respiration is also near the low end of the range of 0.7–7.0 μM C day^{-1} reported for respiration in surface waters [6, 9, 106–109]. As a result, Benner's [4] estimate of an 80 day turnover from a respiration rate of 1 μM C day^{-1} is probably an upper limit for DOC turnover time (Table 1). Higher respiration and other forms of DOC removal (such as aggregation) will combine to shorten this time.

6.1.2
Aggregation and Respiration

The aggregation of larger (>200 nm) colloids by surface coagulation (i.e. the coagulation of the colloids on the surfaces of rising bubbles) triggers short-lived episodes of bacterial respiration and growth that last only a few hours [6]. In the Sargasso Sea, surface coagulation enhances respiration by a factor of 2–6; this will decrease DOC turnover by (i) increasing respiration and (ii) by removing colloidal fractions of the DOC to transfer them up the size spectrum. In order to make any sense of a given respiration measurement, Kepkay [6] has concluded that the episodic response of respiration to colloid aggregation must be accounted for explicitly. Not only should attention be paid to the time scale of data acquisition, but the episodic response of respiration to aggregation should also be incorporated into models of the deep transport of ocean productivity [73, 110].

Given that bioreactive colloids are incorporated into aggregates, the release of ectoenzymes by free-living bacteria [97, 111, 112] and the bacteria associated with aggregates [97–99] is a key step in the mobilization of carbon for respiration. Smith et al. [98] have measured such intense hydrolytic (polymer-degrading) activities that they refer to aggregates as "enzymatic reactors". This concept of aggregates as bioreactors for the respiration of COC is somewhat at odds with the findings of Amon and Benner [7, 8] and Kepkay et al. [9]. They have all produced data suggesting that aggregation is not an absolute prerequisite for a close association to develop between COC and respiration (Fig. 5). However, none of the results to date can be used to definitively rule out aggregation, es-

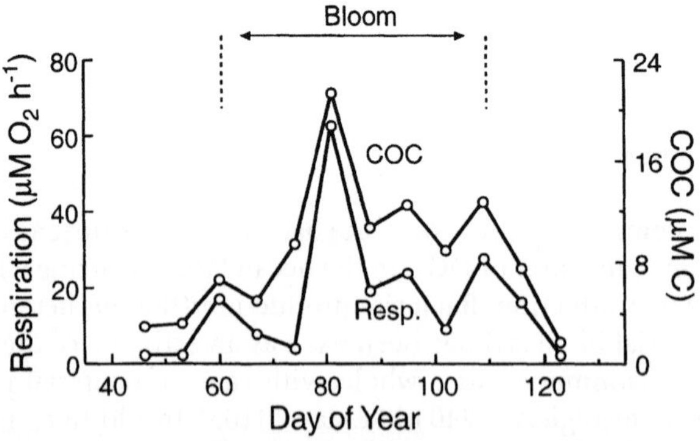

Fig. 5. Association of colloidal organic carbon (COC) and respiration at a depth of 5 m during a spring phytoplankton bloom in Bedford Basin, Nova Scotia [9, 19]

pecially the effects of aggregation on the submicron scale [24]. As a result, any controversy about aggregation being a requirement for the respiration of COC may be more apparent than real and should not obscure the fact that the bacterial utilization of COC can be up to four times higher than the utilization of LMWOC [4].

6.1.3
Accumulation of Bioreactive Carbon

Recently obtained data [113, 114] and a re-examination of older data [115] suggest that DOC accumulates in the photic zone during seasons of high productivity. This can be explained simply by assuming that DOC is not very reactive [116]. However, the accumulation of bioreactive COC during phytoplankton blooms [9, 19] provides an argument against such a simple explanation. During blooms, carbon-rich colloids can accumulate independently of the Redfield carbon to mineral nutrient ratio [9]. In addition, the deep transport of DOC (including colloids) can be regulated by seasonal diffusive and advective mixing rather than by aggregate sinking. On the basis of these two considerations alone, Thingstad et al. [117] have suggested that a biological pump for the deep transport of dissolved organic size fractions would function in a fundamentally different fashion from the classical concept of a pump primed by large sedimenting aggregates.

Thingstad et al. [117] go on to advance the concept of a malfunctioning microbial loop, where bacterial growth in oligotrophic waters is kept low by competition with the phytoplankton for nutrients, and bacterial biomass is kept low by protozoan predation. Under these constraints, DOC consumption cannot match DOC release, allowing bioreactive DOC to accumulate in surface waters. Once the bioreactive carbon has accumulated, it is available for deep transport by seasonal mixing [113, 114] but, in the absence of large aggregate formation [24], it can also remain in the photic zone and be transformed into less reactive material. The slow hydrolysis of polymers [118] has been advanced as a transformation mechanism, but COC can be broken down rapidly [7–9] and slow hydrolysis would be unlikely if ectoenzymes are present. Given the absence of a viable biochemical mechanism, another process – the photochemical transformation of DOC – may be in operation.

6.2
Photoreactivity

DOC is involved primarily in the absorption of the UV component of solar radiation in seawater [119]. As a result, it seems likely that UV radiation is responsible for most of the photochemical transformations of this carbon [120]. The depth to which UV can penetrate in seawater is about 20 m in the case of UV-B (280–320 nm) radiation and about 60 m in the case of UV-A (320–400 nm) radiation [121]. As a result, the photochemical effects of UV are restricted to the upper regions of a photic zone extending to a depth of about 100 m in the open ocean. Photochemical transformations of the organic matter measured as DOC can be caused by the direct absorption of light or indirectly by the production of peroxides and free radicals [122]. The major identifiable photoproduct of DOC is CO_2 [122, 123], but a number of low molecular weight organic products are also produced [100, 101, 120, 122].

6.2.1
Organic Photoproducts – More or Less Bioreactive?

Since the pioneering work of Kieber et al. [100], it has been assumed that organic photoproducts [120] and, in some cases, inorganic products [124, 125] are responsible for the enhancement of bacterial growth after the exposure of DOC to sunlight. There is, however, good evidence that photochemical transformations can produce less reactive as well as more reactive products. For example, Keil and Kirchman [126] have found that bacterial growth on protein extracts is reduced when the extracts are added to estuarine water and irradiated. In addition, Benner and Biddanda [127] have demonstrated that there is a depth dependence in the response of bacteria to the irradiation of natural DOC in seawater, with a 68–84% reduction of growth in irradiated surface water compared with a 22–79% reduction in response to the irradiation of deep water (Fig. 6). Herndl et al. [128] have gone further, pointing out that irradiation can generate more and less reactive DOC at the same time. They have also stressed that the molecular transformations leading to these changes in bioreactivity are not understood because there is an overall lack of detailed information on the effects of UV radiation on bacterial metabolism [129].

6.2.2
Solar Radiation and Bacterial Metabolism

Even though it is clear that bacterial growth in surface waters is inhibited by the effects of solar radiation (Fig. 6), the causes contributing to this decline in growth are less obvious. UV irradiation can result in DNA damage [130] which

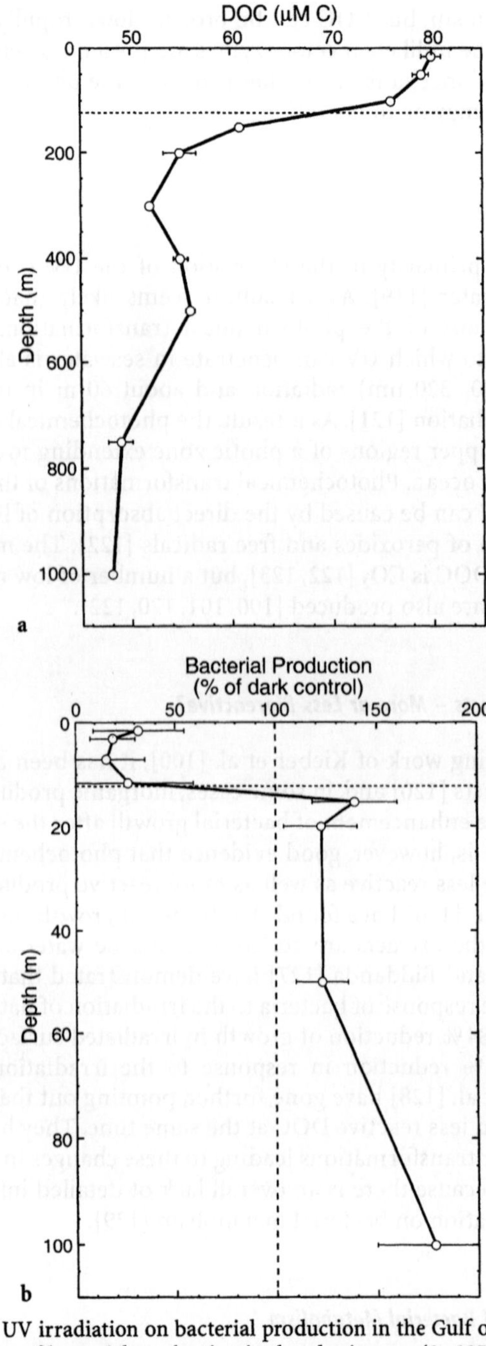

Fig. 6a, b. Effect of UV irradiation on bacterial production in the Gulf of Mexico [127]. Note the strong inhibition of bacterial production in the photic zone (0–125 m), where dissolved organic carbon (DOC) is high. In deeper water, where DOC is low, production is enhanced. The enhancement and inhibition of bacterial production are plotted as percent departures from the production measured in non-irradiated controls (*vertical broken line*)

would obviously tend to inhibit growth, but the degree of inhibition is also related to ambient nutrient concentrations. Kaiser and Herndl [131] have found that nutrient-limited bacteria in coastal surface waters exhibit smaller decreases in growth after exposure to solar irradiation than bacteria growing under nutrient-rich conditions. Even so, bacterial growth can be inhibited by as much as 40% after only 3 h of exposure to near-surface irradiation [131, 132]. Added to this, ectoenzyme activity declines rapidly in response to solar irradiation [133] and, given the role ectoenzymes could play in the degradation of COC [9], this enzymatic inhibition alone would encourage the buildup of bioreactive COC. But perhaps most important of all is the ability of marine bacteria to express photoenzymatic repair mechanisms [131]. Lower irradiation and UV-A at depth in the photic zone induce these photorepair mechanisms and allow the bacteria in surface waters to recover from the damaging effects of solar radiation.

6.2.3
Diurnal Cycling of Bioreactive Carbon

Taken as a whole, the metabolic responses of bacteria to solar irradiation have led Herndl et al. [128] to postulate a diurnal mechanism for the regulation of DOC reactivity. During daylight hours, temporary stratification can extend down to depths of 40 m in the open ocean. Within the stratified region, DOC and planktonic organisms will receive a considerable dose of solar radiation, retarding bacterial growth and generating a mixture of bioreactive and less bioreactive photoproducts. At dusk, stratification breaks down, allowing the bacteria to be mixed deeper. At depth, the bacteria can repair UV-B damage and utilize the bioreactive photoproducts of DOC before the next (daytime) onset of stratification. Repeated episodes of diurnal stratification will result in the buildup of a complex range of photoproducts with a wide range of bioreactivities.

The tendency for more or less bioreactive DOC to build up will not only depend on diurnal cycles of photoinhibition and enzymatic repair, but will also depend on the origin of the DOC. For example, the bioreactive COC released as exudates by phytoplankton will be transformed by solar radiation and accumulate as less bioreactive, low molecular weight, photoproducts [134]. These photoproducts can be utilized once they have been mixed downwards to deeper water [127]. In contrast, the upwelling of deeper water into sunlit surface regions will produce more bioreactive photoproducts from low reactivity DOC [135]. When superimposed on the diurnal effects of solar irradiation on bacterial metabolism, this cycling of bioreactive and less reactive DOC and COC between surface and deep waters will encourage the buildup of old LMWOC at the expense of younger, more reactive COC. At this point, however, the direct inhibitory effects of solar radiation on bacterial metabolism cannot be distinguished from the indirect effect of generating more recalcitrant substrates as photoproducts. The net result of both processes is reduced bacterial growth, but this reduction may be regulated by something as simple as an increase in respiration [129]. This could process a large amount of organic carbon and funnel the energy into cell maintenance rather than growth. As a result, any quantitative index of the bioreactivity of photoproducts will be hard to come by without more detailed studies of the effects of UV radiation on cell metabolism.

7
Summary and Conclusions

Marine scientists tend to view the ocean carbon cycle from vantage points rooted in distinct fields of research. For example, analytical chemists and photochemists tend to focus on carbon analysis, physical chemists on aggregation, and biologists on the reactivity of colloids. So, by taking an integrated overview

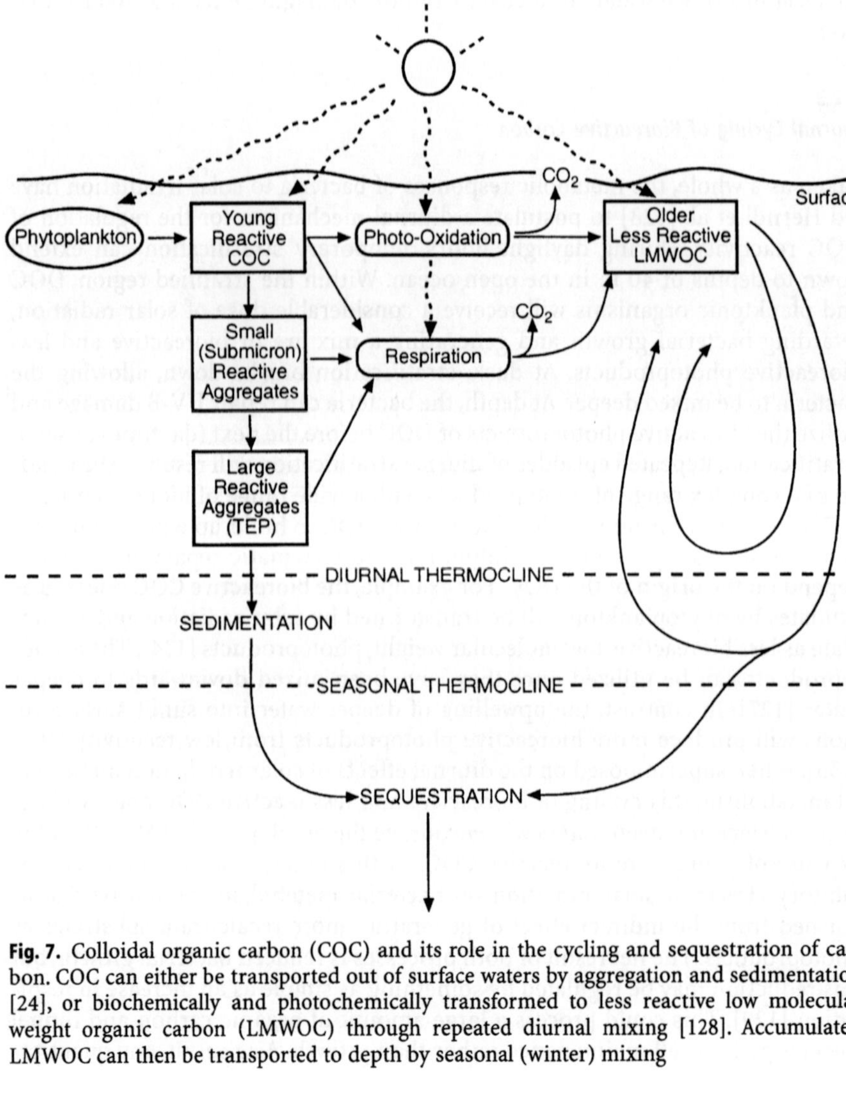

Fig. 7. Colloidal organic carbon (COC) and its role in the cycling and sequestration of carbon. COC can either be transported out of surface waters by aggregation and sedimentation [24], or biochemically and photochemically transformed to less reactive low molecular weight organic carbon (LMWOC) through repeated diurnal mixing [128]. Accumulated LMWOC can then be transported to depth by seasonal (winter) mixing

of the three fields, new insights can be obtained on the role of colloids in the cycling and deep sequestration of carbon:

(1) COC in the surface ocean is a non-living but reactive reservoir of carbon that is globally significant (Fig. 2); it outweighs the living carbon stored in marine biomass by a wide margin. Any accurate model of carbon cycling must take this reservoir into account.
(2) Aggregation (Fig. 4), respiration (Fig. 5) and photooxidation (Fig. 7) are all involved in the removal of COC from the surface ocean. Despite this trio of removal mechanisms, bioreactive COC builds up in surface waters during blooms (Fig. 5). It also accumulates as biochemically and photochemically transformed carbon over extended growing seasons.
(3) The photochemical effects of UV radiation on bacterial metabolism and DOC leads to the buildup of bioreactive and less bioreactive DOC in surface waters. The repeated diurnal cycling of this reactive and less reactive carbon results in the accumulation of old LMWOC at the expense of younger, more reactive COC (Fig. 7).

With respect to the sequestration of carbon in deep water, bioreactive COC is either caught up into aggregates that can sediment out (to fuel the deep ocean ecosystem) or is converted to less reactive LMWOC that can accumulate prior to deep transport by winter mixing (Fig. 7). The degree to which aggregation or accumulation is coupled to the production of DOC is a prime regulator of deep carbon transport. This balance between DOC production and accumulation may, in the words of Thingstad et al. [117], allow deep carbon sequestration to "indeed function very well".

8 References

1. Hedges JI (1992) Mar Chem 39:67
2. Watson AJ, Lovelock JE (1983) Tellus 35B:284
3. Berner RA (1989) Palaeogeogr Palaeoclimatol Palaeoecol 75:97
4. Benner R (1998) In: Hessen H, Tranvik L (eds) Aquatic humic substances. Springer, Berlin Heidelberg New York, p 319
5. Longhurst AL, Sathyendranath S, Platt T, Caverhill C (1995) J Plankton Res 17:1245
6. Kepkay PE (1994) Mar Ecol Prog Ser 109:293
7. Amon RMW, Benner R (1994) Nature 369:549
8. Amon RMW, Benner R (1996) Limnol Oceanogr 41:41
9. Kepkay PE, Jellett JF, Niven SEH (1997) Mar Ecol Prog Ser 150:249
10. Sharp JH, Benner R, Bennett L, Carlson CA, Dow R, Fitzwater SE (1993) Limnol Oceanogr 39:1774
11. Sharp JH, Benner R, Bennett L, Carlson CA, Fitzwater SE, Peltzer ET, Tupas LM (1995) Mar Chem 48:91
12. Carlson DJ, Brann ML, Mague TH, Mayer LM (1985) Mar Chem 16:155
13. Whitehouse BG, Petrick G, Ehrhardt M (1986) Water Res 20:1599
14. Whitehouse BG, Yeats PA, Strain PM (1990) Limnol Oceanogr 35:1368
15. Benner R, Pakulski JD, McCarthy M, Hedges JI, Hatcher PG (1992) Science 255:1561
16. Benner R, Biddanda B, Black B, McCarthy M (1999) Mar Chem in press
17. Ogawa H, Ogura N (1992) Nature 356:696

18. Kepkay PE, Niven SEH, Milligan TH (1993) Mar Ecol Prog Ser 100:233
19. Kepkay PE, Niven SEH, Jellett JF (1997) J Plankton Res 19:369
20. Guo L, Santschi PH, Warnken KW (1995). Limnol Oceanogr 40:1392
21. Buesseler KO, Bauer JE, Chen RF, Eglinton TI, Gustafsson O, Landing W, Mopper K, Moran SB, Santschi PH, Vernon Clark R, Wells ML (1996) Mar Chem 55:1
22. Lerman A (1979) Geochemical processes. Wiley-Interscience, New York
23. Filella M, Buffle J (1993) J Colloid Surf 73:255
24. Buffle J, Wilkinson KJ, Stoll S, Filella M, Zhang J (1998) Environ Sci Technol 32:2887
25. Williams PM, Gordon LI (1970) Deep Sea Res 17:19
26. Druffel ERM, Williams PM, Bauer JE, Ertel JR (1992) J Geophys Res 97:15639
27. Fry B, Sherr EB (1984) Contrib Mar Sci 27:13
28. Hedges JI, Cowie GL, Richey JE, Quay PD, Benner R, Strom M, Forsberg BR (1994) Limnol Oceanogr 39:743
29. Meyers-Schulte KJ, Hedges JI (1986) Nature 321:61
30. Opsahl S, Benner R (1997) Nature 386:480
31. Baines SB, Pace ML (1991) Limnol Oceanogr 36:1078
32. Azam F, Cho BC (1987) In: Fletcher M, Gray RG, Jones JG (eds) Ecology of microbial communities. Cambridge University Press, Cambridge, p 261
33. Suttle CA, Chan AM (1993) Mar Ecol Prog Ser 92:99
34. Strom SL, Benner R, Ziegler S, Dagg MJ (1997) Limnol Oceanogr 42:1364
35. Decho AW (1990) Oceanogr Mar Biol Annu Rev 28:73
36. Koike I, Hara S, Terauchi K, Kogure K (1990) Nature 345:242
37. Wells ML, Goldberg ED (1991) Nature 353:342
38. Tanoue E, Nishiyama S, Kamo M, Tsugita A (1995) Geochim Cosmochim Acta 59:2643
39. Williams PM, Druffel ERM (1987) Nature 330:246
40. Bauer JE, Williams PM, Druffel ERM (1992) Nature 357:667
41. Santschi PH, Guo L, Baskaran M, Trumbore S, Southon J, Bianchi TS, Honeyman B, Cifuentes L (1995) Geochim Cosmochim Acta 59:625
42. Duursma E (1961) Neth J Sea Res 1:1
43. Jackson GA, Williams PM (1985) Deep Sea Res 2:223
44. Hansell DA, Williams PM, Ward BB (1993) Deep Sea Res 40:219
45. Hansell DA, Waterhouse TY (1997) Deep Sea Res 44:843
46. Williams PJleB (1995) Mar Chem 51:17
47. Bishop JKB, Edmond JM, Ketten DR, Bacon MP, Silker WG (1977) Deep Sea Res 24:511
48. Thurman EM (1985) Organic geochemistry of natural waters. Nijhoff/Junk, Dordrecht
49. Buffle J, Perret D, Newman M (1992) In: Buffle J, van Leeuwen HP (eds) Environmental particles, vol 1. Lewis, Boca Raton, p 171
50. Harvey GR, Boran GA (1985) In: Aikin GR, McNight DM, Wershaw RL, MacCarthy P (eds) Humic sustances in soil, water and sediment: geochemistry, isolation and characterization. Wiley-Interscience, New York, p 233
51. Malcolm RL (1990) Anal Chim Acta 232:19
52. Hedges JI, Hatcher PG, Ertel JR, Meyers-Schulte KJ (1992) Geochim Cosmochim Acta 56:1753
53. Clair TA, Sayer BG, Kramer JR, Eaton DR (1996) Hydrobiologia 317:141
54. Gagosian RB, Stuermer DH (1977) Mar Chem 5:605
55. Lee C, Bada JL (1977) Limnol Oceanogr 22:502
56. Johnson KM, Sieburth JMcN (1977) Mar Chem 5:1
57. Pakulski JD, Benner R (1992) Mar Chem 10:55
58. Pakulski JD, Benner R (1994) Limnol Oceanogr 39:930
59. Aluhiware LI, Repeta DJ, Chen RF (1997) Nature 387:166
60. Santschi PH, Balnois E, Wilkinson KJ, Zhang J, Buffle J (1998) Limnol Oceanogr 43:896
61. Mopper K, Schultz CA, Chevolot L, Germain C, Revuelta R, Dawson R (1992) Environ Sci Technol 26:133
62. Mopper K, Zhou J, Ramana KS, Passow U, Dam HG, Drapeau DT (1995) Deep Sea Res II 42:47

63. Borch NH, Kirchman DL (1997) Mar Chem 57:85
64. Skoog A, Benner R (1999) Limnol Oceanogr in press
65. Guo L, Santschi PH (1997) Rev Geophys 35:17
66. Johnson BD, Kepkay PE (1992) Deep Sea Res 39:5
67. von Smolouchowski M (1917) Z Phys Chem 93:129
68. O'Melia CR (1987) In: Stumm W (ed) Aquatic chemical kinetics. Wiley, New York, p 447
69. O'Melia CR (1990) In: Stumm W (ed) Aquatic surface chemistry. Wiley, New York, p 385
70. Lyklema J (1991) Fundamentals of interface and colloid sciences, vols 1 and 2, Academic Press, London
71. Stumm W (1992) Chemistry of the solid-water interface, Wiley-Interscience, New York
72. Stumm W, Morgan JJ (1996) Aquatic chemistry. Chemical equilibria and rates in natural waters, 3rd ed. Wiley, New York
73. Jackson GA, Burd AD (1998) Environ Sci Technol 32:2805
74. Tanford C (1961) Physical chemistry of macromolecules. Wiley, New York
75. Fleet GI, Cohen-Stuart MA, Scheutjens JMHM, Cosgrove T, Vincent B (1993) Polymers at interfaces. Chapman and Hall, London
76. Wilkinson KJ, Ioz-Roland A, Buffle J (1997) Limnol Oceanogr 42:1714
77. Muzzarelli RAA (1973) Natural chelating polymers. Pergamon Press, Oxford
78. Chin W-C, Orellana MV, Verdugo P (1998) Nature 391:568
79. Gordon DC (1970) Deep Sea Res 17:175
80. Alldredge AL, Passow U, Logan BE (1993) Deep Sea Res 40:1131
81. Passow U, Alldredge AL (1994) Mar Ecol Prog Ser 113:185
82. Schuster S, Herndl GJ (1995) Mar Ecol Prog Ser 124:227
83. Mari X, Kiørboe T (1996) J Plankton Res 18:969
84. Mari X, Burd A (1998) Mar Ecol Prog Ser 163:63
85. Mari X (1999) Mar Ecol Prog Ser in press
86. Kiørboe T, Hansen JLS (1993) J Plankton Res 15:993
87. Shimeta J (1993) Limnol Oceanogr 38:456
88. Tranvik CJ, Sherr EB, Sherr BF (1993) Mar Ecol Prog Ser 92:301
89. Flood PR, Deibel D, Morris CC (1992) Nature 355:630
90. Carman KR (1990) Mar Ecol Prog Ser 68:71
91. Decho AW, Moriarty DJW (1990) Limnol Oceanogr 35:1039
92. Schuster S, Arrieta JM, Herndl GJ (1998) Mar Ecol Prog Ser 166:99
93. Niven SEH, Kepkay PE, Bugden JB (1997) Radioprotection Colloques 32:C2–213
94. Long RA, Azam F (1996) Aquatic Microb Ecol 10:213
95. Jackson GA (1995) Deep Sea Res II 42:215
96. Hill PS (1996) Deep Sea Res 43:679
97. Chróst, RJ (1990) In: Overbeck J, Chrost RJ (eds) Aquatic microbial ecology: biochemical and molecular approaches. Springer, Berlin Heidelberg New York, p 47
98. Smith DC, Simon M, Alldredge AL, Azam F (1992) Nature 359:139
99. Hoppe H-G, Ducklow H, Karrasch B (1993) Mar Ecol Prog Ser 93:277
100. Kieber DJ, McDaniel J, Mopper K (1989) Nature 341:637
101. Mopper K, Zhou X, Kieber RJ, Kieber DJ, Sikorski RJ, Jones RD (1991) Nature 353:60
102. Azam F, Hodson RE (1977) Limnol Oceanogr 22:492
103. Ducklow HW, Carlson CA (1992) In: Marshall KC (ed) Advances in microbial ecology, vol 12. Plenum Press, New York, p 113
104. Kirchman DL, Suzuki Y, Garside C, Ducklow HW (1991) Nature 352:612
105. Hansell DA, Bates NR, Gundersen K (1995) Mar Chem 51:201
106. Pomeroy LR, Sheldon JE, Sheldon WM Jr, Peters F (1995) Mar Ecol Prog Ser 117:259
107. Packard TT, Williams PJleB (1981) Oceanol Acta 4:351
108. Smith SV, Hollibaugh JT (1993) Rev Geophys 31:75
109. Biddanda B, Benner R (1999) Deep Sea Res in press
110. Bacasow R, Maier-Reimer E (1991) Global Biogeochem Cycles 5:71
111. Karner M, Herndl GJ (1992) Mar Biol 113:341
112. Christian JR, Karl DM (1995) Limnol Oceanogr 40:1042

113. Copin-Montégut G, Avril B (1993) Deep Sea Res 40:1963
114. Carlson CA, Ducklow HW, Michaels AF (1994) Nature 371:405
115. Williams PJleB (1995) Mar Chem 51:17
116. Legendre L, LeFévre J (1995) Aquat Microb Ecol 9:69
117. Thingstad TF, Hagström Å, Rassoulzadegan F (1997) Limnol Oceanogr 42:398
118. Billen G (1990) Ergeb Limnol 34:191
119. Zika RG (1981) In: Duursma EK, Dawson R (eds) Marine organic chemistry: evolution, composition, interactions and chemistry of organic matter in seawater. Elsevier oceanography series, vol 31, Elsevier, Amsterdam, p 299
120. Moran MA, Zepp RG (1997) Limnol Oceanogr 42:1307
121. Smith RC, Baker KS (1981) Appl Opt 20:177
122. Miller WL, Zepp RG (1995) Geophys Res Lett 22:417
123. Granéli W, Lindell M, Tranvik LJ (1996) Limnol Oceanogr 41:698
124. Francko DA, Heath RT (1982) Limnol Oceanogr 27:564
125. Bushaw KL, et al. (1996) Nature 381:404
126. Keil RG, Kirchman DL (1994) Mar Chem 45:187
127. Benner R, Biddanda B (1999) Limnol Oceanogr in press
128. Herndl GJ, Arrieta JM, Kaiser E, Obernosterer I, Pausz C, Reitner B (1999) In: Proceedings of the 8th international symposium on microbial ecology, Halifax, Nova Scotia (Aug 9–14, 1999)
129. Benner R, Ziegler S (1999) In: Proceedings of the 8th international symposium on microbial ecology, Halifax, Nova Scotia (Aug 9–14, 1999)
130. Jeffrey GJ, Pledger RJ, Aas P, Hager S, Coffin RB, von Haven R, Mitchell DL (1996) Mar Ecol Prog Ser 137:283
131. Kaiser E, Herndl GJ (1997) Appl Environ Microbiol 63:4026
132. Herndl GJ, Müller-Niklas G, Frick J (1993) Nature 361:717
133. Müller-Niklas G, Heissenberger A, Puskaric S, Herndl GJ (1995) Aquat Microb Ecol 9:111
134. Pausz C, Herndl GJ (1999) Aquat Microb Ecol in press
135. Lindell MJ, Granéli W, Tranvik LJ (1995) Limnol Oceanogr 40:195
136. Eppley RW, Harrison WG, Chisholm SW, Stewart E (1977) J Mar Res 35:671
137. Cauwet G (1978) Oceanol Acta 1:99
138. Siegenthaler U, Sarmiento JL (1993) Nature 365:119
139. Stewart JE (1999) Fisheries and Oceans Canada, personal communication based on NAFO catch statistics

CHAPTER 3

Gas Exchange at the Sea Surface

Bruce D. Johnson

Oceanography Department, Dalhousie University, Halifax, Nova Scotia, B3H 4J1, Canada
E-mail: bjohnson@is.dal.ca

An accurate model describing gas exchange across the air-sea interface is essential: for assessing the role of the oceans in the global cycles of many environmentally important gases; for determining the strength of biological activity through monitoring dissolved O_2; and for interpreting results in many studies in which dissolved gases are used as tracers of marine processes. Results of measurements of gas exchange suggest that wind speed is not sufficient as a predictor of gas exchange rate and that other variables such as fetch, wave parameters, surface active materials and dynamics of bubbles may need to be considered, as well. More data are needed from measurements at sea under well-characterized conditions in order to identify the minimum subset of variables that must be measured in order to accurately predict gas exchange rates. New methods for measuring gas concentrations using moored instruments promise to provide data that cover gas exchange through the full range of natural oceanic and meteorological conditions and in so doing facilitate further model development.

Keywords: Air-water gas exchange, Gas fluxes, Gas transfer, Dissolved gases.

The Handbook of Environmental Chemistry Vol. 5 Part D
Marine Chemistry
(ed. by P. Wangersky)

1 Introduction

The oceans serve as a source or sink for many gases of environmental importance. For example, gases such as CO_2, N_2O, and CH_4 are of concern as greenhouse gases, methyl halides are implicated in the destruction of the ozone layer, dimethyl sulfide is important as a source of cloud condensation nuclei, and ozone and CO are tropospheric pollutants. Determining ocean uptake and production of these and other environmentally important gases is essential for constructing the global budgets that are necessary for assessing the extent to which human activities are altering natural cycles.

However, the interest in gases in the oceans goes beyond understanding their bulk production and uptake, for the distributions of dissolved gases offer a wealth of information about biological, chemical, geological, and physical processes. Seasonal cycles of O_2 are used to estimate rates of photosynthesis (e.g. [1]). Dissolved ^3He in the deep ocean provides information on the rate of outgassing of the earth's interior and flow through ocean vents [2]. Gases that include chlorofluorocarbons (freons), CCl_4, $^{13}CO_2$, tritium, ^3He and $^{14}CO_2$ are used to study mixing processes and water mass transport (e.g. [3–6]), and the distributions of various other dissolved gases offer evidence for the importance of particular chemical and biological pathways (e.g. [7–9]).

In many studies involving dissolved gases, interpretation of measurements requires an understanding of the rate of gas transfer across the air-water interface. For example, determining the strength of the oceans as a source or sink for trace gases usually involves measuring the concentrations of the gas on both sides of the air-sea interface and then estimating the ocean flux through application of empirical relationships that express transfer rate as a function of wind speed [10]. If it is assumed that the atmosphere and surface ocean are in steady state with regard to the gas, or if measurements are averaged over long time periods, the calculated flux provides a local estimate of the strength of the ocean as a source or sink.

Rates of gas transfer must also be determined in studies using O_2 for estimating the net rate of biological production. In this type of study, deviations from O_2 saturation in the euphotic zone are interpreted in terms of the net result of biological activity (production minus respiration), transport across the thermocline, horizontal advection, and gas transfer through the air-sea interface [1]. Measurement of inert gases in addition to O_2 allows the results to be separated into contributions from biological and physical processes [11–15]. Where measurements are integrated over year-long time-scales, net production can provide an estimate of export of carbon to the deep ocean.

In the use of dissolved gases as tracers for studying mixing processes in the ocean, often it is sufficient to assume that the water mass was in equilibrium with the atmospheric concentration of the gas at the time that the water was last in contact with the atmosphere. However, under more dynamic conditions, where equilibration may not occur, e.g. in regions of strong vertical exchange [16] or for gases that have a source in situ, an understanding of gas exchange rates is needed for interpretation of results (e.g. [4]).

2
Gas Exchange Fundamentals

2.1
Resistances to Gas Transfer

The rate at which gases pass between the atmosphere and ocean mixed layer is determined by the rate of transfer through the thin diffusion-dominated boundary layer at the gas-water interface. However, the state of this interface and the transfer of gases across it are dependent on wind, breaking waves, bubbles, Langmuir circulation, turbulence, water and air temperatures, humidity and adsorbed surface active materials (Fig. 1). The contribution of all of these to gas exchange is the focus of current research; the goal is to determine the minimum subset of measurements needed to accurately predict the rate of exchange of gases over the natural range of meteorological and oceanographic conditions.

For gases passing between the atmosphere and ocean, two resistances are important (Fig. 2). These resistances are the boundary layers on the air and water sides of the interface. At some distance from the interface on both sides, the mediums are well mixed. As the interface is approached, turbulent mixing is damped by viscous forces and, very near the interface, diffusion becomes important for mass transfer. The resistance to transfer of any gas is the sum of the air side and the water side resistance. For gases that are highly soluble or react with water, e.g. SO_2, NH_3 and H_2O vapor, the water side resistance becomes less important and the air side resistance dominates. In this case, it is passage through

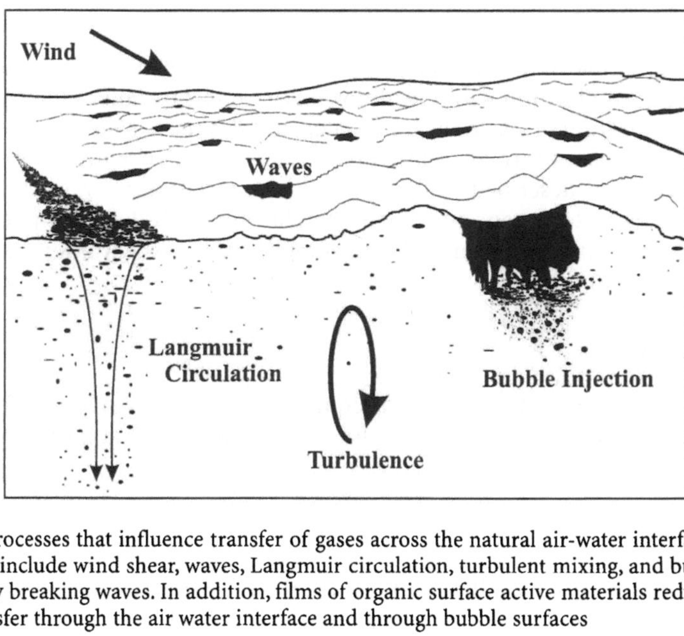

Fig. 1. Processes that influence transfer of gases across the natural air-water interface. Some of these include wind shear, waves, Langmuir circulation, turbulent mixing, and bubbles injected by breaking waves. In addition, films of organic surface active materials reduce direct gas transfer through the air water interface and through bubble surfaces

the air side boundary layer that determines the rate of gas exchange. Conversely, most of the gases that do not react with water, such as the principal atmospheric gases, N_2, O_2 and Ar, encounter greater resistance on the water side and passage through the boundary layer on the water side becomes rate determining.

Only a few gases, e.g. formaldehyde, PCBs and DDT [10], pass through both diffusive boundary layers at comparable rates and, thus, both resistances become important for these gases. Carbon dioxide is also a special case, because CO_2 reacts with water, but at a rate that is sufficiently slow (about 1 min) that at all but the lowest wind velocities its transfer is not enhanced by the reaction. While the reaction of CO_2 with water can be catalyzed by the naturally occurring enzyme carbonic anhydrase, measurements of the enzyme in bulk seawater have not found the concentration to be high enough to significantly enhance CO_2 transfer rates [17]. However, carbonic anhydrase may be enriched in the surface micro-layer, but this possibility has not yet been investigated [18].

2.2 Piston Velocity

Rates of gas exchange are expressed in terms of the piston velocity, also called the exchange velocity, transfer velocity, or exchange coefficient. The piston velocity is the rate of exchange of a gas per unit area, per unit difference in concentration across the interface (driving force) and is the same as the mass transfer coefficient used by engineers. Practically, the piston velocity is the proportionality constant, K_w, that relates gas flux to difference in concentrations across the air-water interface or

$$F = K_w (C_{air} - C_{water}) \tag{1}$$

in which C_{air} is the concentration of gas at the top of the interface, i.e. the concentration at equilibrium with the atmospheric concentration of the gas. C_{water} is the concentration of gas in the bulk water, and F is the gas flux, i.e. the rate at which the gas passes through a unit area of interface. Conceptually, the piston velocity has been described as the net rate of translation of two pistons, one pushing the gas, at the concentration in equilibrium with the atmosphere, into the ocean and the other pushing the gas at bulk water concentration out [3].

2.3 Gas Solubility

In order to determine the state of disequilibrium and hence driving force for gas exchange, it is necessary to know the solubility of the gas at the temperature and salinity of the water phase.

$$C_{(T,S)} = k\,p \tag{2}$$

in which k is the gas solubility, and $C_{(T,S)}$ is the concentration of the gas at the temperature and salinity of the water parcel at equilibrium with the partial pressure, p, of the gas in the gas phase. While many different terms are used to express solubilities, most commonly, solubility is given as the dimensionless Henry's law constant (the ratio of the concentration in the gas phase to the concentration in un-ionized form in the water at equilibrium); the Bunsen absorption coefficient (volume of the gas at standard temperature and pressure in a unit volume of solution at equilibrium with one atmosphere of the gas in the gas phase); or in moles or standard cubic centimeters of the gas per kilogram of solution in equilibrium with one atmosphere of the gas. While solubility is also a function of ambient pressure, this dependence is negligible over the natural range of pressures at the sea surface [19]. The solubilities of many gases have been determined as a function of temperature and salinity, and compilations of results and descriptions of experimental methods appear in the literature. For example, see the many publications of Weiss and co-workers (e.g. [20–23]) and others [24–26].

3 Models of Gas Exchange

Three levels of models have been used to describe gas exchange:

- conceptual models that are descriptive but not predictive because they include unknown or adjustable parameters,
- theoretical models based on particular fundamental processes such as turbulence, wave formation, bubble injection or white capping, and
- empirical models derived from experimental results and field observations

3.1 Conceptual Models

As the surface between the air and water is approached from either side, turbulence is damped and advective transport of gases to and away from the interface is reduced. The boundary condition often applied requires that the perpendicular fluid velocity at the interface be zero.

Considerable discussion has appeared in the literature over the thickness of the layer at the air-water interface through which diffusion is important [27]. One way in which this layer thickness has been estimated is through use of a simple model called the stagnant film model (Fig. 2). In this model the flux,

$$F = (D/z)\,(\Delta C) \tag{3}$$

in which D, the gas diffusivity in water, divided by a diffusion layer thickness, z, is the piston velocity. Based on measurements of gas exchange, estimates of z for gases with predominant water-side resistance (normal gases) range from a little more than 10 µm to hundreds of microns [3].

However, the nature of these estimates needs to be considered in terms of the processes involved. At very low wind speeds, the thickness of the diffusive region influenced by proximity to the interface must extend some distance into

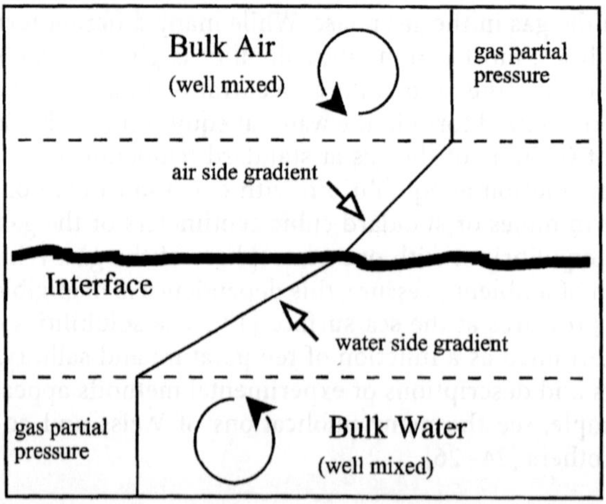

Fig. 2. Schematic showing resistances to gas transfer at the air-water interface. Gases that are highly soluble, e.g. as the result of reacting with water, experience dominant resistance on the air side, whereas gases that are sparingly soluble, including most atmospheric gases, encounter greater resistance on the water side. In the stagnant film model, the resistance is due to diffusion through a stagnant layer that produces a linear gradient as shown in the diagram. Note that if the resistances on the air and water sides of the interface were equal, the diffusion layer on the air side would be about D_{air}/D_{water} or about 10^4 times thicker than the water side diffusion layer

the water phase. Mixing from below erodes the base of this diffusion layer but in general is not sufficient to overcome surface tension forces and sweep the interface into the bulk water. As wind speed or mixing becomes more energetic, the effective diffusive layer is thinned, the interface between gas and water is periodically renewed as turbulent eddies sweep the surface layer into the bulk water phase, and the interface undergoes expansion and contraction by wave action. At still higher wind speeds, additional interfacial area is created through formation of bubbles and spray droplets.

For low wind speeds, one might ask, at what characteristic depth or distance from the interface is the rate of advective transfer of normal dissolved gases away from the interface equal to diffusive transfer? To answer this question, we can use the dimensionless Peclet number, (dV/D), which expresses the relative importance of mass transfer by advection to transfer by diffusion. In the Peclet number, d can be taken as the thickness of the diffusive layer, V the velocity and D as the gas diffusivity in the water phase. If we take V as the piston velocity with an appropriately low value of about 1 cm h^{-1}, and a typical diffusivity for gases in water of about 10^{-5} cm^2 s^{-1}, the thickness of the boundary layer can be determined for a Peclet number, Pe = 1, i.e. at a distance from the interface where advective and diffusive transport are comparable[1]. Under these conditions, d is

[1] While this calculation provides a sense of the effective thickness of the boundary layer, conceptually it is the same as assuming transport through a stagnant film. At low wind speeds gas transfer likely occurs through diffusion into the boundary layer from the gas phase and then transport due to erosion of the boundary layer from below by small turbulence eddies.

predicted to be about 350 μm. Note that the calculated thickness is a function of diffusivity and, thus, its magnitude depends on the gas under consideration.

The meaning of estimates of the diffusion layer thickness at higher wind speeds is less clear. Under more dynamic conditions, capillary and larger waves extend the interface, bubbles enhance gas transfer and turbulence renews the interface. Thus, any estimate of diffusion layer thickness based on measurements of gas transfer rates must be considered only a nominal or effective thickness. In this regard, Broecker and Peng [3] estimate a global average thickness of the diffusion layer of about 40 μm based on rates of invasion of natural radiocarbon $^{14}CO_2$.

The Peclet number also gives a means for comparing measurements of gas exchange for gases with different physical properties, specifically different diffusivities in water, and at different temperatures and salinities. The nature of fluid flow is characterized by another dimensionless number, the Reynolds number, Re, or $(dV\varrho/\mu)$. Re, in which d is a characteristic distance, V is velocity, ϱ is water density and μ is fluid viscosity, gives an indication of the importance of inertial forces relative to viscous forces. Dividing the Peclet number by the Reynolds number gives $\mu/\varrho D$, the dimensionless Schmidt number, Sc. Usually, μ/ϱ is expressed as the kinematic viscosity, ν, and the Sc as ν/D. For gas transfer, exchange rates have been generally accepted as varying as $(Sc)^{-2/3}$ at low wind speeds and $(Sc)^{-1/2}$ at moderate and higher wind speeds. The $-2/3$ dependence is predicted for a surface approximated as a smooth rigid wall [28], and the $-1/2$ dependence is for a free surface without adsorbed surfactant. While the value of the exponent has not yet been established from oceanic measurements, Watson et al. [29] found a dependence of $(-1/2)$ for gas exchange for lakes. Evidence from wind-wave facilities suggests a transition from $-2/3$ to $-1/2$ as the mean square slope of the waves increases [30].

Measurements of piston velocities for different gases and at different temperatures and salinities are often compared for a Schmidt number of 600, i.e. the Schmidt number for CO_2 at 20 °C. However, the influence of bubbles, spray and adsorbed surfactants may substantially alter the dependence of gas exchange on the Sc [18], and thus scaling of different gas exchange measurements based on assumptions about the value of the exponent may be sometimes in error.

The stagnant film model (Eq. (1)) provides a clear conceptual model of gas exchange, but appears to have little support from experimental results. This is because the stagnant film model predicts a dependence of exchange on D to the first power, Sc^{-1}, a dependence that is not consistent with results of measurements.

Another simple model, the surface renewal model [31], predicts a $D^{1/2}$ dependence. In this model, the interface between the air and water is renewed periodically by turbulence eddies, a process that mixes gas that has diffused into the water surface down into the bulk phase (Fig. 3). Jacobs [32] gives the quantity of substance diffusing across a plane between two semi-infinite mediums as:

$$Q_{0,t} = 2\,Au_0\,(Dt/\pi)^{1/2} \tag{4}$$

in which A is the interfacial area, t is the time of diffusion, and u_0 is the concentration. If we take the plane of interest as the air-water interface, and ΔC the

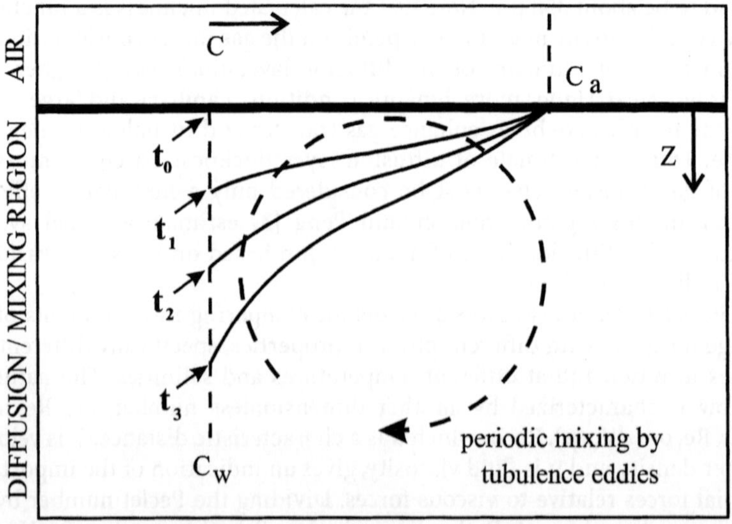

Fig. 3. Schematic of the surface renewal model of gas exchange (shown, water side resistance dominant). In this model, gas diffuses across the air-water interface and establishes a gradient. Turbulence eddies periodically mix the gradient into the bulk water phase. *Solid lines* show gas concentration with distance from the interface and correspond to times, t_0 to t_3, which represent increasing times between renewal events. The concentration of dissolved gas is C, and subscripts "a" and "w" refer to air (at equilibrium) and water, respectively. Note that the shorter the surface renewal time, the higher the gas transfer, because transfer occurs on average along a stronger gradient

difference in concentration between the bulk water and water at equilibrium with the gas phase, Eq. (2) can be divided by a characteristic surface renewal time to give the gas flux:

$$F = (Q/At_s) = 2\ (D/t_s\ \pi)^{1/2}\ \Delta C \tag{5}$$

In this model the piston velocity is equal to $2\ (D/t_s\ \pi)^{1/2}$. The surface renewal time, t_s, is not known a priori and is usually calculated from data. While the surface renewal model is conceptually more attractive than the stagnant film model, it too has an adjustable parameter, t_s, and thus is also not predictive.

3.2 Models Based on Fundamental Processes

The effect of turbulence in creating new surface is implicit in the surface renewal time of Eq. (5), but t_s is not easily determined from measurable turbulence parameters. A number of more explicit models have been developed. These differ in how the interaction of turbulent eddies with the interface is parameterized [28, 33–36].

Many theoretical and experimental treatments of gas transfer parameterize exchange in terms of the friction velocity, u^*, rather than wind speed, because hydrodynamic characteristics of the interfacial region are more directly related to the friction velocity. For example, the thicknesses of the viscous boundary layers on the air and water sides of the interface are proportional to the respective friction velocities. The friction velocity is defined as $u^* = (\tau/\rho_a)^{0.5}$, in which τ is the wind stress and ρ_a is air density. The magnitude of u^* in a particular experimental setting is sometimes determined directly from measurements, but when such measurements are not available, u^* is determined from the wind speed using an empirical drag coefficient, i.e. from $\tau = C_D \rho_a U_{10}^2$ in which C_D is the drag coefficient and U_{10} is wind speed at 10 m elevation.

Deacon [28] developed a boundary layer model based on turbulent fluid flow in the vicinity of a smooth rigid wall. By assuming that the wind stress is continuous across the air-water interface, producing a constant flux of momentum, the friction velocity on the water side can be determined as $u^*_w = u^*_a(\rho_a/\rho_w)^{0.5}$ in which "a" and "w" refer to air and water, respectively. This approach has been found to provide a reasonable description of gas transfer in wind tunnels at low wind speeds [10]. Another boundary layer model [35] allows some surface divergence and predicts the $-2/3$ power of the Sc for low wind speeds and $-1/2$ power at higher wind speeds.

Csanady [37] has argued for the importance of surface divergence associated with rollers on breaking wavelets for maintaining a thin diffusional boundary layer. The model predicts a $Sc^{-1/2}$ dependence for gas transfer. Monahan and Spillane [38] have proposed a model in which bubbles produced by breaking waves enhance turbulence in the near-surface and, thus, ocean whitecaps provide a low resistance vent for gas exchange. This model has been used by Erickson [39] to estimate global transfer rate patterns for ocean uptake of anthropogenic CO_2.

A number of modeling studies have treated the effects of bubbles in gas transfer [40–47]. From echo-sounder measurements, Kanwisher [48] found that bubble clouds were transported to depths of 10 m or more under storm conditions and from these observations suggested that bubbles are important for gas exchange. Thorpe [41] provided a detailed analysis of the contribution of bubbles based on experimental observations and modeling, and concluded that bubbles in the sea should significantly enhance gas transfer for wind speeds of 12 $m s^{-1}$ and may dominate at higher wind speeds. Results of other modeling studies have shown that gas supersaturations of up to a few percent should occur at steady state because surface tension and hydrostatic pressures are always positive and cause bubble dissolution [42, 43, 45]. The asymmetry of the gas exchange process that results from bubble dissolution should be greatest for gases with lower solubilities [43]. Other modeling results show that larger bubbles in breaking waves are important for gas transfer and should also influence transfer of more soluble gases such as CO_2 [47].

Studies of the effect of surface active materials on gas exchange have shown that natural surfactants present in seawater have a first-order effect on gas transfer [49–51]; see reviews by Frew [52] and Asher [53]. In general, direct inhibition due to added resistance of adsorbed films does not appear to be important for limiting gas transfer in natural waters but, rather, hydrodynamic effects involving changes in near surface turbulence, and damping of waves are probably the principal mechanisms by which surface active materials retard gas transfer. The effect of surfactant films on turbulence is to alter length and velocity scales of turbulence eddies that interact with the interface [52]; surface renewal is suppressed and the concentration gradient near the interface expands [51, 54]. Under conditions of strong suppression of surface renewal, the surface behaves as a solid wall and the piston velocity becomes proportional to $D^{2/3}$ as in the model of Deacon [28, 52].

Gas transfer through the air-water interface is also sensitive to other processes that alter the interface, e.g. processes such as condensation, evaporation and heat transfer. Typically, strong temperature gradients exist in the top millimeter of the sea surface [55]. The so-called skin temperature is estimated globally to be about 0.3 °C cooler than the bulk water temperature [56, 57]. Consideration of the skin temperature in estimating global CO_2 absorption by the oceans provides an additional 0.6 giga-tons of uptake, which is an amount that helps reconcile model estimates of the ocean sink for anthropogenic CO_2 with estimates based on gas transfer considerations [18, 57]. However, McNeil and Merlivat [58] point out that another process, warming of the top few meters of the surface ocean by solar radiation, can counter the thermal skin effect, increasing CO_2 evasion and decreasing CO_2 invasion.

3.3 Empirical Models

Figure 4 shows some of the results of measurements of gas exchange in fresh water lakes and in the surface ocean, expressed in terms of piston velocity as a function of wind speed, U_{10}, i.e. wind speed measured at 10 m above the water surface. From measurements such as these, and results from wind tunnel experiments, empirical models of piston velocity as a function of wind speed have

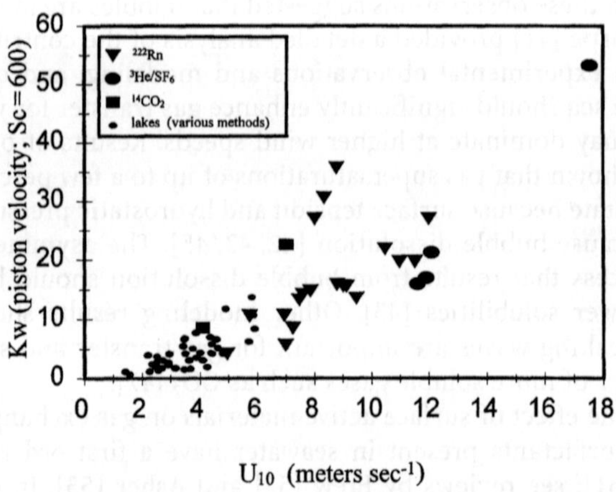

Fig. 4. Some field measurements of piston velocity as a function of wind speed, U_{10}, the wind speed measured at 10 m elevation above the water surface. Data are taken from [18, 26, 52] which have compiled results from the original sources

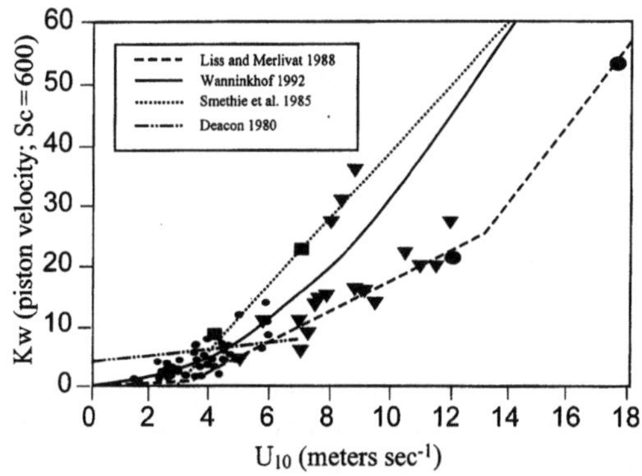

Fig. 5. Data from field measurements as in Fig. 4, but with some of the empirical relationships that have been proposed (see text for details)

been proposed. One of the more widely used empirical formulations was presented by Liss and Merlivat [59] and includes three linear regimes that correspond to changes in mass transfer processes: at wind speeds below 3.6 m s^{-1} capillary waves have not yet appeared, and the water surface is very nearly smooth; between wind speeds of 3.6 and 13 m s^{-1}, capillary and larger waves appear; at wind speeds above 13 m s^{-1} breaking waves inject bubbles which further increase the rate of gas transfer with wind speed. A number of data sets support this relationship; however, there are few data for gas transfer in the bubble-mediated regime and, especially, at the high wind speeds associated with storms. Results from a dual tracer experiment [29] that were obtained under storm conditions fall very nearly on the proposed Liss and Merlivat line. However, other field results [60, 61] and modeling results [16] suggest that the predictions of Liss and Merlivat underestimate the importance of bubbles in the exchange process.

A number of other functions have been proposed that relate piston velocity to wind speed. These are shown along with the Liss and Merlivat [59] relationship in Fig. 5. On the basis of TTO data from measurements in the Equatorial Atlantic using the radon deficiency method, Smethie et al. [62] proposed a linear relationship between wind speed and gas exchange. Wanninkhof [26] proposed a quadratic function and noted the effect of using an average wind speed when the dependence of exchange on wind speed is nonlinear. The relationship is for steady winds or short-term winds and was determined from the data for bomb $^{14}CO_2$ and the wind speed distribution around the global mean wind speed. The model predictions of Deacon [28], also shown in Fig. 5, are appropriate only to low wind speeds.

4 Methods

A broad range of methods has been used to study gas exchange in natural waters. Most methods require collection of water samples, extraction of the gas of interest and then analysis in shipboard or shore-based laboratories. This experimental regimen is difficult at sea and particularly at higher wind speeds. New methods that provide measurements by unattended instruments, e.g. attached to moorings, are very much needed, and will stimulate rapid gains in model testing and development.

4.1 Methods in Use

The Dual Tracer Method. This involves releasing into the surface layer known amounts of a gaseous tracer substance and a second tracer for following mixing. Ideally, the second tracer should be nonvolatile, detectable at low concentrations, and not naturally present. In practice, finding an appropriate tracer for mixing has been difficult. The inert substance SF_6 has been used and is often coupled with the gas ^3He. Because SF_6 is also volatile, the ratio of exchange rates is obtained and, thus, determination of piston velocity requires assumptions about the dependence of exchange on the Schmidt number. Usually $Sc^{-1/2}$ is assumed. However, when bubbles mediate exchange, the Sc number exponent may differ from (–1/2), but the form of the dependence is not yet known [26].

An advantage of the dual tracer method is that the exchange rate is determined on time scales that are comparable to changes in wind forcing. Some of the disadvantages include: there is not at present an obvious choice for a tracer to follow mixing; the method can be used to measure evasion only; and water sampling and measurements on shipboard in a rough sea are difficult and can introduce error.

The Radon Deficiency Method. This is based on determining the radioactive disequilibrium between ^{222}Rn and its precursor, ^{226}Ra [63]. Away from the air-sea and sediment-water interfaces, the two radioisotopes are generally in decay equilibrium. Deficiency of ^{222}Rn activity in the surface mixed layer is due to loss of the gas to the atmosphere, and from this disequilibrium between nuclides a transfer velocity can be calculated. One requirement of this method is that winds be steady for several days – a condition that often is not met at sea. In addition, the method measures evasion rates only.

The CO_2 Eddy Flux Method. This method provides the only direct measure of gas exchange. Used in agriculture and forest research, where fluxes are about an order of magnitude greater than exchange through the sea surface, the method is based on measuring fluctuations in vertical wind velocity and CO_2 concentration. In results reported to date, instrument noise has been a concern [64–66]. The next generation of sensors, however, promises at least an order of magnitude improvement in noise immunity. Another problem with the method, as currently used, has been the necessity of deploying the instrument on a stable platform, but the possibility exists for airplane or ship-mounted use.

The $^{14}CO_2$ methods. These use both natural and bomb-produced ^{14}C to infer exchange rates [67]. In the natural $^{14}CO_2$ method, exchange of the gas across the air-sea interface is assumed to be in balance with loss of the isotope through radioactive decay in the ocean. A second use of $^{14}CO_2$ utilizes the input to the atmosphere from nuclear testing. Excess $^{14}CO_2$ due to invasion of bomb-produced $^{14}CO_2$ has been measured in profiles in many parts of the world oceans. This excess is interpreted in terms of the known $^{14}CO_2$ distribution in the atmosphere since nuclear testing began.

Other Inventory Methods. Time series of dissolved gases have been collected [11, 13, 68] and used to elucidate processes such as gas exchange, mixing, new production and the contributions of temperature changes and bubbles to gas supersaturation. Through measuring dissolved O_2 and inert gases with a range of physical properties (diffusivity and solubility), the strength of physical and biological processes can be determined. Measurements of argon are especially useful, because Ar is an inert gas with physical properties (diffusivity and solubility) similar to those of O_2.

Separation of biological and physical contributions to changes in O_2 concentration in the ocean surface layer has also been carried out by measuring nutrients. Redfield [69] used changes in phosphorus concentration to quantify the effect of biological processes on O_2 concentration. This approach is based on the now classical work [70] demonstrating that during photosynthesis 138 moles of O_2 are released for each mole of phosphorus taken up. In respiration, i.e. decomposition of plant material, the ratio is again 138:1 for O_2 consumed to phosphorous released (note that some variation in the Redfield ratio has been observed).

Measurements of O_2 combined with measurements of other substances such as phosphorus or inert gases that allow separation of biological and physical contributions to observed changes provide a powerful means of studying gas exchange, because biological processes can produce relatively large deviations from saturation for O_2. However, many studies of O_2 have failed to effectively elucidate the dependence of exchange rate on physical forcing, because sampling frequencies have not provided the time resolution needed to determine fluxes of O_2 under more energetic conditions.

One study that did feature nearly continuous measurement of O_2 is that of Wallace and Wirick [60], who deployed O_2 sensors on a mooring in the N. Atlantic Bight. Data obtained during the winter showed that storm events produced high supersaturations of O_2 on time scales of hours and that, during more quiescent periods, diffusion across the air-sea interface and community biological respiration caused gas saturations to be slowly reduced. This work demonstrated that injection of bubbles substantially enhances gas invasion rates – a conclusion that follows from their observation that wave parameters are better than wind speed as a predictor of gas exchange rate.

4.2 New Methods

The work of Wallace and Wirick [60] shows the value of continuous measurements of dissolved gases using unattended gas sensors in situ. One promising new method that utilizes measurements in situ involves pressure measurements behind a rigidly supported gas permeable membrane to determine gas tension (the sum of the partial pressures of all dissolved gases) [61, 71, 72]. In one study, four of these gas tension devices (GTDs) were deployed at 5 m intervals beneath the sea surface in the Strait of Georgia and provided a detailed time series of gas invasion during passage of a storm [61].

Several obvious advantages exist in measuring gas tension. First, pressure can be measured very accurately with highly stable pressure sensors. Second, measurements of O_2 and gas tension allow an estimate of the partial pressure of N_2 to be made. This estimate is based on the fact that under conditions typical of natural waters, O_2, N_2 and water vapor comprise more than 99% of the gas tension signal. Thus, subtraction of the O_2 partial pressure and water vapor pressure (practically, a function of temperature and salinity) from gas tension provides very nearly the partial pressure of N_2, a gas that is considered almost inert in the oceans.

An instrument now being tested in our laboratory selectively and quantitatively absorbs specific gases from behind the membrane in the gas tension instrument. When O_2 is removed and a correction applied for water vapor, the resulting pressure can be compared with gas tension, and the partial pressure of O_2 can then be determined by difference. This method (gas tension/absorption method) provides both O_2 and N_2 partial pressures and thus can be used for net production estimates and for determination of gas exchange rates. The stability of the gas tension sensor (an accuracy of about 0.2% yr^{-1}) makes it a promising candidate for remote deployment on moorings to give long-term, unattended measurements covering a broad range of oceanographic and meteorological conditions.

The gas tension/absorption method can also be used to measure other gases, such as CO_2, by changing the specific gas absorbent in the sensor. Other in situ sensors for measuring CO_2 have also recently been developed [73–75] and are based on principles such as infrared (IR) CO_2 analysis and fluorescence quenching of a specific dye by CO_2. These new methods are being tested at sea.

Recent studies have shown the importance of measurements of wave parameters [60, 76] and sea surface (skin) temperature for determination of piston velocity. These physical measurements of the sea surface are now available from satellite remote sensing data and can be used along with in situ and other measurements of gas partial pressures or concentrations in the ocean surface layer to determine fluxes. Indeed it has long been an objective of the climate change community to seed critical regions of the world oceans with sensors for measuring CO_2 and then to determine fluxes of the gas through estimation of piston velocities using satellite data. Through use of these methods, the flux of CO_2 through the sea surface could be used to provide large-scale estimates of uptake of CO_2 produced by human practices. Such methods for determining CO_2 fluxes are very much needed to assess the current role of the oceans in the global carbon cycle and to provide a capability for long-term monitoring of CO_2 fluxes in a changing climate.

While progress is being made in developing stable in situ gas sensors, new methods are being developed for studying gas exchange directly at the air-water interface. One of these, the controlled flux method [76], provides information about local fluxes on time scales of minutes. In this method, a controllable tracer flux is applied directly to the interface. When steady state is achieved, i.e. when the flux is constant, the transfer velocity is determined from the difference between the concentration of the tracer at the water surface and in the bulk. While, theoretically, the method can be used with gases, in practice, heat flux has been measured and related to gas transfer [76]. For heat flux measurements, an IR source is used to heat the water surface, and the surface temperature is measured using an IR detector. The method promises to provide much new information on processes that contribute to direct gas exchange at low and moderate wind speeds.

5
Measurements of Gas Exchange in Natural Waters

In an assessment of our understanding of uptake of anthropogenic CO_2 by the oceans, a recent report of progress of a major international climate change program (the Joint Global Ocean Flux Study, JGOFS) terms gas exchange, "not well understood", and lists this understanding as no better than 50%. An impediment to development and testing of models has been the lack of agreement in measurements of gas exchange in natural waters. While much has been learned from measurements of gas exchange in wind-wave tanks [30], the limited dimensions of these facilities impose difficulties in interpretation of results [77].

The lack of agreement in results, such as those in Fig. 4, has stimulated considerable discussion in the gas exchange community about reasons for observed differences (e.g. [26]). While wind speed is routinely measured by ships, met buoys, and at shore stations, and can be estimated from satellite scatterometer measurements, it now appears clear that wind speed alone is not sufficient as a predictor of gas exchange and that other variables such as fetch, surface active materials, bubbles, and time and space scales of measurements are all important.

Wanninkhof [26] assessed the importance of fetch and variability of wind speed as contributors to differences in gas transfer results. Fetch, the distance over water that the wind blows, is important because wave fields develop over distances of hundreds of kilometers and, thus, gas exchange measurements in wind-wave facilities, and in lakes and at fetch-limited coastal sites, may be different from those in well-developed seas at open-ocean sites. The curvature in relationships that express gas exchange rates as a function of wind speed can also explain some differences in measured gas exchange rates. As Wanninkhof [26] points out, an average wind speed derived from measurements of variable winds will tend to overestimate the effect of wind speed on gas exchange and place points on the curve above those obtained from measurements of exchange for short-term steady winds. However, upon assessing effects of wind speed variability and differences in fetch among reported measurements, Wanninkhof [26] concluded that much of the variability still appears to be due to other causes.

Analysis of data from different wind-wave facilities has shown that gas exchange is substantially increased with the appearance of waves. Jähne et al. [30] showed that wave parameters are a reliable predictor of gas transfer rates; they specifically found a strong correlation between transfer rate and the mean square slope of waves. In field studies, Wallace and Wirick [60] determined that transfer rates correlate better with the significant wave height, the average height of the largest 1/3 of waves, than with wind speed.

Another source of variability in measurements is due to natural surface active materials that result from biological and chemical processes. Frew et al. [50] found in laboratory studies that surfactant materials from various phytoplankton species could decrease O_2 invasion rates by 5–50%. Because of breaking waves and turbulence, intact surface slicks exist only to wind speeds as high as about 6 m s^{-1}, but a surface excess of soluble surface active materials is maintained at considerably higher wind speeds and will influence gas exchange rates accordingly [52, 53].

Variability in bubble populations and dynamics of formation and mass transfer must also contribute to variability in reported gas transfer velocities (for a recent review see [78]). At higher wind speeds, bubbles strongly mediate gas transfer [41, 45, 60, 61]. Bubble populations produced by breaking waves are substantially different in fresh water and seawater [79] and are also likely to vary depending on water temperature, atmospheric pressure, and the presence of surface active matter. Slauenwhite and Johnson [80] found that populations of bubbles produced by break-up of a 5 μl bubble in passage through a small orifice increased by a factor of 3–5 in number in seawater relative to fresh water. They also found that lower temperatures and the presence of natural surface active materials from a diatom bloom significantly enhanced bubble production.

The effect of surface active materials on bubble mass transfer must also be important. There is little doubt that small bubbles become coated with surface active matter and rise and exchange gases at rates below those for bubbles that are "clean" [81–84]. Johnson and Slauenwhite (unpublished data) found that mass transfer of gases from bubbles in a diatom culture medium occurred at rates substantially below those for bubbles in water that had been treated to remove most organic matter. In some cases, measured rates of dissolution of bubbles rising in the culture medium were less than rates predicted for pure diffusion alone.

The effect of surface active matter on mass transfer from bubbles may substantially affect measured gas exchange rates at higher wind speeds. While the accumulation of surfactants at the bulk air-water interface is reduced under more dynamic conditions, collection of dissolved, colloidal and larger particulate matter on bubbles and the effect of these adsorbed materials on gas transfer can be expected to increase under more dynamic conditions. Bubbles are injected to greater depths at higher wind speeds and, thus, surface films accumulate and age over longer bubble rise times. In addition, bubbles assume a greater role in gas transfer at higher wind speeds and, consequently, any variability in bubble mass transfer will especially affect gas exchange under more dynamic conditions.

6
References

1. Jenkins WJ, Goldman JC (1985) J Mar Res 43:465
2. Jenkins WJ, Clarke WB (1976) Deep Sea Res 23:481
3. Broecker WS, Peng T-H (1982) Tracers in the sea, 1st edn. Lamont-Doherty Geological Observatory, Columbia University, New York
4. Jenkins WJ (1988) Phil Trans R Soc Lond A 325:43
5. Quay PD, Tilbrook B, Wong CS (1992) Science 256:74
6. Wallace DWR, Beining P, Putzka A (1994) J Geophys Res 99:7803
7. Zafiriou OC, McFarland M (1981) J Geophys Res 86:3173
8. Keiber RJ, Zhou X, Mopper K (1990) Limnol Oceanogr 35:1503
9. Moore RM, Groszko W, Niven SJ (1996) J Geophys Res 101:28, 529
10. Liss PS (1983) Gas transfer: experiments and geochemical implications. In: Liss PS, Slinn WGN (eds) Air-sea exchange of gases and particles. D. Reidel, Dordrecht, Holland, p 241
11. Craig H, Hayward T (1987) Science 235:199
12. Emerson S (1987) J Geophys Res 92:6535
13. Spitzer WS, Jenkins WJ (1989) J Mar Res 47:169
14. Oudot C (1989) J Mar Res 47:385
15. Emerson S, Quay P, Stump C, Wilbur D, Knox M (1991) Global Biogeochem Cycles 5:49
16. Sarmiento JL, Orr JC, Siegenthaler U (1992) J Geophys Res 97:3621
17. Goldman JC, Dennett MR (1983) Science 220:199
18. Liss PS, Watson AJ, Brock EJ, Jahne B, Asher WE, Frew NM, Hasse L, Korenowski GM, Merlivat L, Philiips LF, Schluessel P, Woolfe DK (1997) Report of group I – physical processes in the microlayer and the air-sea exchange of trace gases. In: Liss PS, Duce RA (eds) The sea surface and global change. Cambridge University Press, Cambridge, p1
19. Andrews FC (1972) Science 178:1199
20. Weiss RF (1970) Deep Sea Res 17:721
21. Weiss RF (1971) J Chem Engin Data 16:235
22. Weiss, RF (1974) Mar Chem 2:203
23. Warner MJ, Weiss RF (1985) Deep Sea Res 32:1485
24. MacKay D, Shiu WY, Sutherland RP (1979) J Am Chem Soc 13:333
25. Moore RM, Geen CE, Tait VK (1995) Chemosphere 30:1183
26. Wanninkhof R (1992) J Geophys Res 97:7373
27. Horne RA (1969) Marine chemistry, 1st edn. Wiley-Interscience, New York
28. Deacon EL (1977) Tellus 29:363
29. Watson AJ, Upstill-Goddard RC, Liss PS (1991)Nature 349:145
30. Jähne B, Münnich KO, Bösinger R, Dutzi A, Huber W, Libner P (1987) J Geophys Res 92:1937
31. Higbie R (1935) Am Inst Chem Eng 35:365
32. Jacobs MH (1967) Diffusion processes. Springer, Berlin Heidelberg New York
33. Fortesque GF, Pearson JRA (1967) Chem Eng Sci 22:1163
34. Lamont JC, Scott DS (1970) Am Inst Chem Eng J 16:513
35. Ledwell JR (1984) The variation of gas transfer coefficient with moelcular diffusivity. In: Brutsaert W, Jirka GH (eds) Gas transfer at water surfaces. D. Reidel, Dordrecht, Holland, p 293
36. Brumley BH, Jirka GH (1988) Physicochem Hydrodynam 10:295
37. Csanady GT (1990) J Geophys Res 95:749
38. Monahan EC, Spillane MC (1984) The role of oceanic whitecaps in air-sea gas exchange. In: Butsasert W, Jirka GH (eds) Gas transfer at water surfaces. D. Reidel, Dordrecht, Holland, p 495
39. Erickson DJ III (1993) J Geophys Res 98:8471
40. Atkinson LP (1973) J Geophys Res 78:962
41. Thorpe SA (1982) Phil Trans R Soc Lond A 304:155
42. Merlivat L, Memery L (1983) J Geophys Res 88:707

43. Memery L, Merlivat L (1985) Tellus 37B:272
44. Kerman BR (1984) J Geophys Res 89:1439
45. Woolfe DK, Thorpe SA (1991) J Mar Res 49:435
46. Woolfe DK (1993) Atmosphere-Ocean 31:517
47. Keeling RF (1993) J Mar Res 51:237
48. Kanwisher J (1963) Deep Sea Res 10:195
49. Goldman JC, Dennett MR, Frew NM (1988) Deep Sea Res 35:1953
50. Frew NM, Goldman JC, Dennett MR, Johnson AS (1990) J Geophys Res 95:3337
51. Asher WE, Pankow JF (1986) Tellus 38B:305
52. Frew NM (1997) The role of organic films in gas exchange. In: Liss PS, Duce RA (eds) The sea surface and global change. Cambridge University Press, Cambridge, p 121
53. Asher, W 1997. The sea-surface microlayer and its effect on global air-sea gas transfer. In: Liss PS, Duce RA (eds) The sea surface and global change. Cambridge University Press, Cambridge, p 251
54. Asher WE, Pankow JF (1991) The effect of surface films on concentration fluctuations close to a gas/liquid interface. In: Wilhelms SE, Gulliver JS (eds) Air-water mass transfer. American Society of Civil Engineers, New York, p 68
55. Ewing GC, McAllister ED (1960) Science 131:1374
56. Robertson JE, Watson AJ (1992) Nature 358:738
57. Van Scoy KA, Morris KP, Robertson JE, Watson AJ (1995) Global Biogeochem Cycles 9:253
58. McNeil CL, Merlivat L (1996) Geophys Res Let 23:3575
59. Liss PS, Merlivat L (1986) Air-sea gas exchange rates: introduction and synthesis. In: Buat-Menard P (ed) The role of air-sea exchange in geochemical cycling. D. Reidel, Dordrecht, Holland, p 113
60. Wallace DWR, Wirick CD (1992) Nature 356:694
61. Farmer DM, McNeil CL, Johnson BD (1993) Nature 361:620
62. Smethie WM, Takahashi TT, Chipman DW, Ledwell JR (1985) J Geophys Res 90:7005
63. Broecker WS, Peng T-H (1971) Earth Planet Sci Lett 11:99
64. Smith SD, Jones EP (1985) J Geophys Res 90:869
65. Smith SD, Jones EP (1986) J Geophys Res 91:10529
66. Smith SD, Anderson RJ, Jones EP, Desjardins RL, Moore RM, Hertzman O, Johnson BD (1991) J Geophys Res 96:8881
67. Broecker WS, Peng T-H (1974) Tellus 24:21
68. Schudlich R, Emerson S (1995) Deep Sea Res 43:569
69. Redfield AC (1948) J Mar Res 7:347
70. Redfield AC (1934) James Johnston memorial volume. Liverpool University Press
71. Anderson ML, Johnson BD (1992) J Geophys Res 97:17899
72. McNeil CL, Johnson BD, Farmer DM (1995) Deep Sea Res 42:819
73. Friederich GE, Brewer PG, Herlien R, Chavez F (1995) Deep Sea Res 42:1175
74. Merlivat L, Brault P (1995) Sea Technology 10:23
75. DeGrandpre MD, Hammar TR, Wallace DWR, Wirick CD (1997) Limnol Oceanogr 42:21
76. Jähne B, Libner P, Fischer R, Billen T, Plate EJ (1989) Tellus 41B:177
77. Hasse L (1990) Tellus 42B:250
78. Woolfe DK (1997) Bubbles and their role in gas exchange. In: Liss PS, Duce RA (eds) The sea surface and global change. Cambridge University Press, Cambridge, p 173
79. Monahan EC, Zeitlow CR (1969) J Geophys Res 74:6961
80. Slauenwhite DE, Johnson BD (1999) J Geophys Res 104:3265
81. Tedesco R, Blanchard DC (1979) J Rech Atmos 13:215
82. Detsch RM (1990) J Geophys Res 95:9765
83. Harris IA, Detsch RM (1991) J Geophys Res 96:8907
84. Detsch RM (1991) J Geophys Res 96:8901

CHAPTER 4

Marine Pollution

Vladimir Zitko

St. Andrews Biological Station, 531 Brandy Cove Road, St. Andrews, NB, E5B 2L9, Canada
E-mail: zitkov@mar.dfo-mpo.gc.ca

The chapter is an introduction to topics of interest in marine pollution. It describes the changes that have occurred during the past 40 years, sources of data, general techniques of quality control, data evaluation, and determination of temporal trends. The status of knowledge of classes of substances of interest, ranging from simple inorganics and organics to complex effluents, is presented. The measurement techniques have improved significantly and the amount of information about the pollution of the marine environment has very much increased. However, the list of substances of interest is not getting shorter and there is a continuing need for faster and cheaper methods of data acquisition, evaluation, information extraction, and integration. All cause and effect relationships may never be fully understood or predicted and there is an urgent need to limit the amount of waste reaching the marine environment.

Keywords: Pollution, Metals, Hydrocarbons, Pesticides, Complex effluents.

The Handbook of Environmental Chemistry Vol. 5 Part D
Marine Chemistry
(ed. by P. Wangersky)

1 Marine Pollution

This chapter is a brief overview of current topics in marine pollution and is intended as an introduction to the field. It is not an exhaustive review and chemical, toxicological, and ecological methods are not discussed, as these are available from many monographs or manuals (see for example [1, 2]). References are selected primarily to introduce sources of information, to illustrate the topics of marine pollution, current state of the knowledge, and to provide some predictions of future developments. A somewhat less chemically-oriented review has been published recently [3].

Increased interest in marine pollution goes back about 40 years, roughly coinciding with the publication of Rachel Carson's Silent Spring and with one of the first major oil tankers accidents, the Torrey Canyon episode. It is interesting to read some of the earlier publications on the subject of marine pollution, to obtain a picture of the situation then and of the perceived and predicted problems, to compare these with the current situation, and, perhaps, to wonder what will people think of the work, ideas, and actions of the 1990s 40 years from now.

The effect of a chemical on the environment (open ocean, coastal waters, estuary, aquatic fauna, aquatic flora, etc.) depends on the toxicity of the chemical and on the amount of the chemical the environment is exposed to (for example, the amount of chemical discharged, the administered dose, the concentration of chemical, and the length of exposure). Accordingly, to determine the effects, two factors have to be investigated: toxicity and exposure. Identification of hazards requires studies of toxicity, whereas exposure data are needed for the estimation of risk. Risk is the probability that the exposure conditions are such that the hazards may materialize and result in an effect. Toxicity data consist of dose-response relationships and, in the aquatic environment, dose is usually given by the concentration and the length of exposure.

An "inverse" situation arises when there are observed effects but unknown causes. Only seldom and mostly in cases of very dramatic effects are the causes unequivocally determined. The identification of the Amnesic Shellfish Toxin, domoic acid, as the active principle of toxic mussels is one example [4]. More often there are just indications of causal relationships, as for example in the common seal mortality in the North Sea [5]. As can be seen from Figs 1 and 2, dead and surviving seals may be reasonably well distinguished by their levels and profiles of PCBs. It is known now that the primary cause was a virus, but the possibility of immunosuppression by organochlorine compounds as a participating factor has not been ruled out [6]. Problems involved in studies of these "inverse" situations have been discussed by Sindermann [7]. Because of the difficulties in establishing effect and cause relationships, preventative or remedial actions should not be delayed until science provides a definitive proof of a relationship, since, as mentioned above, the latter can seldom happen. This is the "precautionary principle" [8]. At the same time, an over-reaction may do more harm than good. One should strive for a compromise solution which considers alternatives, technical feasibility, and cost. In any case, better understanding of the marine environment, natural changes, and fluctuations will help but will not guarantee a separation of the effects of natural processes from those resulting from anthropogenic causes.

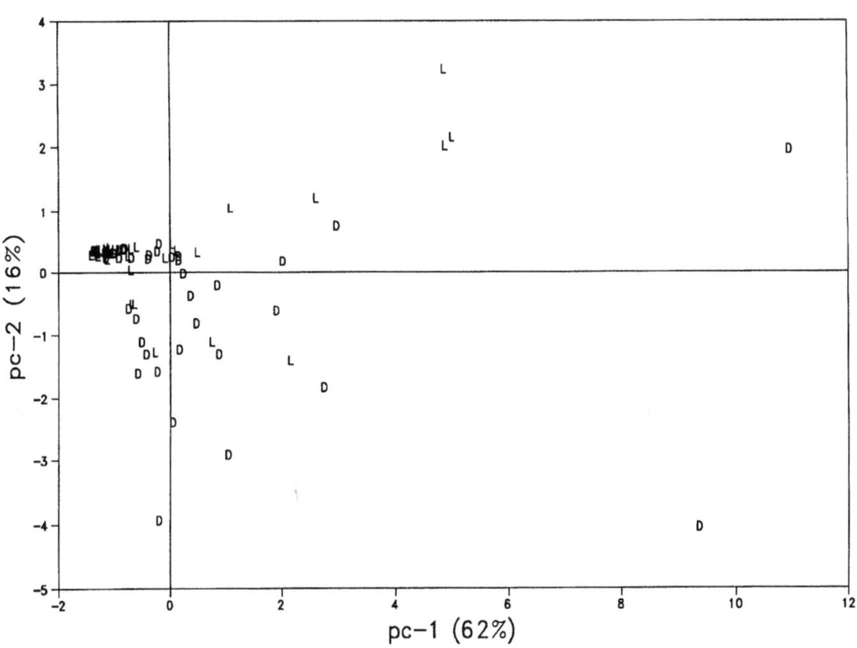

Fig. 1. PCBs in common seals after the phocine distemper epizootic, projected on the plane of the first two principal components. The portions of original variance accounted for are indicated on the axes. L = live, D = dead seals. Calculated from the data of Hall et al. (1992) [6]

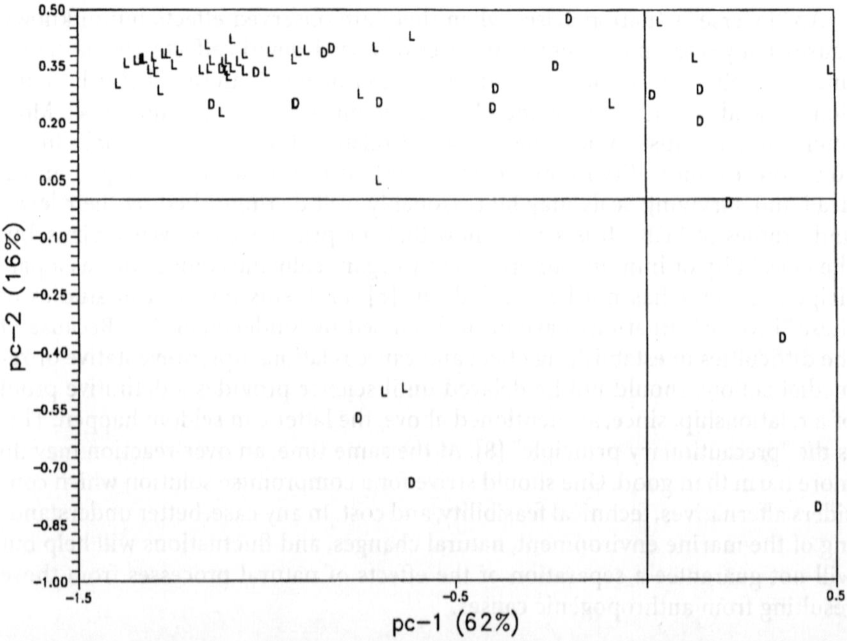

Fig. 2. PCBs in common seals after the phocine distemper epizootic, projected on the plane of the first two principal components; enlarged section of Fig. 1 The portions of original variance accounted for are indicated on the axes. L = live, D = dead seals. Calculated from the data of Hall et al. (1992) [6]

2 Topics of Interest

In the "early" days, biology was the leading discipline in marine pollution studies and the generally recognized effects of marine pollution were "the fishless estuary, the oiled seabirds, and the murky waters carrying an unmentionable scum" [9]. Since the oceans were widely used for waste disposal, of primary concern was the determination of toxicity of the various wastes discharged into the marine environment. The substances of concern included organochlorine and organophosphate pesticides, oil and oil dispersants, mercury, other heavy metals, radioactive waste, and sewage. Additional research needs were anticipated in relation to the continuing exploitation of marine resources. Traditional fisheries were expected to require better fishing techniques and population models maximizing yields without depleting stocks. Aquaculture was predicted to expand from sedentary to other species and to new coastal areas, and research was expected to deal with aquaculture wastes and the need to develop "aquaculture" pesticides. The "drugs from the sea" activity was anticipated to increase. The extent of the extraction of non-renewable resources from the seabed, both in terms of materials extracted and areas affected, was receiving attention. It was recognized that the identity of pollutants discharged into the environment must be known, that chemical industry will create additional problems in the future, and that there is a need, in current terminology, for hazard identification and risk assessment. It was also realized that there must be a better understanding of dispersion processes and that new waste treatments must be developed. The dangers of drawing conclusions about environmental effects and environmental quality objectives from short-term laboratory toxicity tests were recognized.

Petroleum, nutrients (nitrogen and phosphorus), pulp-mill effluents, mercury, lead, chlorinated hydrocarbon pesticides (DDT, the "aldrin/toxaphene" group, and BHC), industrial chlorinated hydrocarbons (PCBs, 1,2-dichloroethane, freons, dry-cleaning solvents), microbes, and radioactivity were identified as pollutants of concern by another group of scientists [10]. The highest priority was given to research on PCBs. Detailed studies of the properties of petroleum films on the sea surface, and more data on the levels of heavy metals in seafood, were called for. Additional recommendations were provided for studies of dispersal and transport, bioavailability, bioconcentration, target tissues, food chain accumulation, models, different levels of toxicity tests (cells, organisms, communities, populations), and studies of degradation and ultimate fate.

The issues related to the exploitation of the oceans have changed little over the almost 40 years gone by. The list of substances of concern grew longer, mainly thanks to the increased environmental awareness, better understanding of environmental behavior of chemicals, and more sensitive analytical techniques. At the same time, practically all substances singled out 30 years ago are still on the list. An example of the application of environmental chemistry is an attempt to group organochlorine compounds according to their physical and chemical properties (molecular weight, vapor pressure, water solubility, octanol/water partition coefficient (K_{ow}), sediment adsorption coefficient), toxicity, persistence, production, and use patterns [11]. The decisions made on the basis of these data are of necessity quite arbitrary. As a very preliminary assessment, GESAMP decided to designate as potentially harmful to the marine environment organochlorine compounds (other than PCBs, PCDDs, and pesticides) with $\log(K_{ow}) > 3$, persistence > 1 week, and toxicity (LC50, EC50) < 10 mg/l. According to these criteria, three groups of organochlorine compounds were formed: chemicals meeting at least two of the above criteria (or similar to those that meet the criteria), chemicals that do not, and chemicals for which data are not available [11].

Chemical and physical properties and some toxicological "benchmarks" must form the basis of an assessment. Other factors, such as dispersion, degradation rates, and similarity to studied compounds, are much more qualitative and expert judgment is required. Models predicting environmental concentrations are useful in the process. A major problem is the fact that single pure chemicals are seldom if ever discharged into the environment. Chemicals are almost always encountered as complex mixtures in effluents, leachates, or formulations. Quite often the presence of some chemicals may not even be anticipated. Groups of experts with different scientific backgrounds and strong chemical analytical support will remain the best tool for the selection, hazard identification, risk assessment, and risk management of chemicals [12].

A more recent prediction of issues emerging in the next 5–15 years [13] identifies municipal discharge, resource extraction, storm sewer, agricultural (pesticides and fertilizers) runoff, land disposal, sedimentation, erosion, overfishing, and toxic contamination of aquatic organisms. Estuaries and coastal water are the sensitive areas, and non-point source pollution the difficult-to-control problem. The latest recognized additional topics of concern are algal blooms, related to increased concentrations of nutrients in coastal waters and to the dissemination of non-indigenous organisms by ballast waters, and environmental oestrogens [14].

On the whole, the major issues in the next 10 years are global change, pollution, waste, biodiversity, natural and man-made risks and hazards, and sustainable management of resources [15]. It may be that the recent simultaneous collapse of many major fisheries can be classified as global change. All of these issues are ultimately related to population growth and limits of the environment to accept all the waste products. From "resource limits" of the 1970s, environmental problems have changed to "recipient limits" [16].

Among the chemicals that are only now slowly attracting attention are nitroaromatic compounds and fluorinated organics. Nitrobenzene in the ng/l range and 1-chloro-2-nitrobenzene in one order of magnitude lower concentrations are widely distributed in the German Bight, and musk xylene (1-*tert*-butyl-3,5-dimethyl-2,4,6-trinitrobenzene) and musk ketone (1-*tert*-butyl-3,5-dimethyl-2,6-dinitro-4-acetyl-benzene) have also been detected in a somewhat lower concentration range [17]. Many fluorinated organics have biological activity and are persistent in the environment [18].

3
Sources of Data

A considerable amount of data had been assembled by 1976 on chlorinated hydrocarbons, petroleum hydrocarbons, heavy metals, and radioactivity [19]. For the first time, littering of the marine environment, primarily by plastics, was identified as an emerging problem. Early toxicity data for both marine and freshwater environments were summarized as well [20].

More recent data may be found in a number of journals publishing papers on marine pollution. Two of the journals most widely read and dedicated to the marine environment are the Marine Pollution Bulletin and Marine Environmental Research, but all of the large number of journals dealing with the environment contain many papers on marine pollution. Marine Pollution Research Titles, published by the Plymouth Marine Laboratory, is an excellent source of titles and also covers the "grey" literature very well. In addition, various reports on the state of the marine environment are published periodically. Two such reports were issued by GESAMP, the most recent being [21]. In addition there are periodical reports dealing with specific areas such as the Baltic [22, 23] and the Scottish [24] Seas. Other reports appear occasionally [25–30]. The concentrations of organochlorine pesticides, PCBs and PAHs in bivalves from Central and South America have been reported by the International Mussel Watch (IMW) Project [31].

Somewhat less visible are large sources of data assembled by the International Council for the Exploration of the Sea (ICES). ICES Advisory Committee on the Marine Environment issues annual reports that summarize advice on various marine environmental topics in response to requests from other agencies, as well as on topics considered important by the committee (see for example [32]). The annexes included in the reports are usually background scientific papers that may eventually appear in primary literature. Other ICES reports review environmental studies, such as monitoring programs [33], or summarize the results of intercalibration exercises [34]. The amount of data accumulated by ICES on contaminants in biota is remarkable. There are some 275,000 records on contaminants in invertebrates, fish, birds, and mammals, about 280,000 records on contaminants in sea-water, about 80,000 records on contaminants in sediments, and 4000 records on biological effects [32]. The data span a period of 20 years and are gradually being made available on the World Wide Web.

A recent review of aquatic sediments [35] contains many references to pollutants in marine sediments. The United States National Oceanic and Atmospheric Administration (NOAA) issues reports on the concentration of chemicals in sediments (see for example [36, 37]). The United States Environmental Protection Agency has a similar program dealing with estuaries [38]. The results of a 1990/1991 study of contaminants in North Sea [39] and Baltic [40] sediments have been summarized. In the UK, a program on the interaction of land and sea, the Land Ocean Interaction Study (LOIS) was initiated in 1993, and, on an international scale, within the International Geosphere Biosphere Programme, the Land-Ocean Interactions in the Coastal Zone (LOICZ), was started in the same year [41]. Among the objectives of these programs is better understanding of biogeochemical processes and the transport and fate of pollutants in estuaries and the coastal zone. First results of LOIS have been published in a special volume [41].

A review of concentrations of mercury, cadmium, lead, PCBs, DDT and metabolites, and HCH in water, sediments, and biota in coastal as well as open ocean areas of the world has been published [42].

Over 70 estuarine and marine macrobenthic organisms, mostly native to Europe and North America, have been used to date in sediment toxicity studies [43] and there are at least 7 other potentially useful species. The commonly used endpoints are, in increasing order of sensitivity, mortality, development, growth, behavior, and, for some chemicals, accumulation. Unfortunately, there are no established tests for mutagenicity.

Other publications deal with the effects and fate of pollutants and experimental techniques for such studies (see for example [44]). In addition, the effects of pollutants on marine organisms are periodically reviewed (for the last review in the series by Reish et al. see [45]). Similar periodic reviews are published for mixing and transport processes [46]. Toxicity data for marine fauna are also available from the U.S. Environmental Protection Agency database AQUIRE. Generally speaking, the sensitivity of marine fauna to chemicals varies widely, particularly among the crustaceans [47]. The International Maritime Organisation (IMO) has a database of toxicity and environmental effect data on Hazardous Substances Carried by Ships [48].

Concentrations of contaminants in sediments, obtained by the NOAA and EPA programs, have recently been evaluated for the ability to predict the toxicity of sediments from the concentrations of contaminants [49]. The study shows that the contaminant concentrations serve as indicators of the need for further study but they should not be used as the only criterion of toxicity.

The probability that a sediment is highly toxic in an amphipod survival test (survival in test significantly lower than in control) when all chemicals (8 metals, 13 PAHs, PCBs, DDTs, dieldrin, and lindane) are below their low effect range (ERL) is 9%. When the concentration of one or more chemicals is above the ERL, but none exceeds the median effect range (ERM), the probability is 17% in the amphipod test and 60–64% in any one other standard test (sea urchin tests, Microtox tests). When the concentration of one or more chemicals is above the ERM, the probabilities are 50% and 86%, respectively [38]. It would be interesting to evaluate the sediment composition and toxicity data by multivariate chemometrical and statistical techniques.

In another study, a 10-day survival test with the amphipod *Ampelisca abdita*, a 48-h survival and development test with bivalve *Mulinia lateralis* embryos exposed to sediment elutriates, and a Microtox test on methylene chloride extracts were used. Chemicals measured in the sediments included metals, PAHs, and the usual chlorinated pesticides and PCBs. For most of the contaminants the mean concentrations in toxic and non-toxic samples did not differ very much. This indicates that many contaminants contribute to the toxicity. PAHs were most consistently correlated with the toxicity [50].

Another approach to the evaluation of sediment quality is the Sediment Quality Triad assessment [51, 52]. This assessment combines three components: concentrations of chemicals, laboratory toxicity tests, and benthic fauna surveys. Each of the components may consist of a number of measurements. The triads may be presented graphically in various formats or subjected to multivariate analysis.

An increasing amount of information may be obtained via the World Wide Web from organizations such as the International Maritime Organization (IMO), the Organization for Economic Cooperation and Development (OECD), the International Atomic Energy Agency (IAEA), the United Nations Environment Program (UNEP), and many other sites. Information from official international agencies is generally objective and accurate. Critical evaluation may be required when dealing with the information presented by others on the Internet.

Not related directly to marine pollution, but important for gaining an insight into potential developments in chemical technology, the composition of formulations, new chemicals, and new uses of existing chemicals, are chemical journals such as the Chemical Engineering News and CHEMTECH, published by the American Chemical Society.

4 Data Quality Control

Some of the early data were affected by the lack of good analytical methods, instrumentation, and, particularly for heavy metals in sea-water and for PCBs in general, by contamination. Over the years, methods and instrumentation have improved remarkably, many intercalibration exercises have been carried out, and quality control and assurance procedures have been implemented [53]. However, the Horwitz observation that relative standard deviation doubles when the concentration decreases by two orders of magnitude is still valid [54]. The resulting errors may be considerable. For example, for organic pollutants in water, analyzed by five laboratories, the coefficient of variation was 500% and the number of laboratories reporting the concentration of benzene in water within 40% of the true concentration decreased from 80% to 15% at benzene concentrations of 7.5 µg/l and 0.12 µg/l, respectively [55].

The quality of the chemical analyses is improving and the control and assurance procedures are being increasingly formalized and expanded worldwide. These activities are organized by a program entitled "Quality Assurance of Information for Marine Environmental Monitoring in Europe" (QUASIMEME), which now includes 126 participating laboratories [56]. The chemicals covered are nutrients, metals, chlorobiphenyls and organochlorine pesticides, PCDD and PCDF, PAHs, pentachlorophenol, triazines, and organophosphates. The matrices are brackish water and sea-water, sediment, and biota. A related program is "Quality Assurance of Sampling and Sample Handling" (QUASH), initiated in 1997 [57]. The program will deal with the effects of sampling and sample handling in the measurement of nutrients in sea-water, of lipid, water, chlorobiphenyls, and metals in biota, and of size fractionation (< 63 micron and < 20 micron), organic carbon, heavy metals, aluminum, lithium, PCBs, and PAHs in sediments.

The International Mussel Watch (IMW) project covered in its initial phase South and Central America. Several species of bivalves have been collected since no single species is common to all sampling areas. Samples were analyzed by two laboratories for organochlorine pesticides and PCBs. One laboratory also measured PAHs. It is unfortunate that in the intercalibration, there were two- to four-fold differences in measured concentrations between the two laboratories.

5 Data Evaluation

In addition to the usual statistical methods based on univariate descriptors (mean, median, and standard deviation) and analysis of variance, multivariate techniques of statistics and chemometrics are increasingly being used in data evaluation. Whereas the former are more rigorous in theoretical background and assumptions, the latter are useful in the presentation of the data, pattern recognition, and multivariate calibrations. Several good monographs on chemometrics are available (see for example [58–61]).

Chemometrical techniques, particularly Principal Component Analysis (PCA) are very well suited for the evaluation of data on multicomponent mixtures such as PCBs, TCDDs, TCDFs, or PAHs. The reduction of the dimensionality of the data leads to two-dimensional or three-dimensional projections which facilitate pattern recognition and interpretation of the results. PCA is used mainly for pattern recognition such as the distinction between control and contaminated samples, changes in profiles of mixtures of compounds such as PAHs or PCBs, identification of outliers, etc. (see for example [62–67]).

PCA was able to distinguish three sources of PCDD and PCDF in crab hepatopancreas: pentachlorophenol, polychlorophenol condensation during pulp digestion, and chlorine bleaching [68]. Following the introduction of controls on the discharge of PCDD and PCDF by the pulp mills in British Columbia (Canada), the concentrations of the chlorine bleaching-related tetrachloro PCDFs have decreased more rapidly than the concentrations of hexachloro PCDDs. The latter are associated mainly with polychlorophenol condensation products and with pentachlorophenol.

The applications of PCA are not limited to multicomponent mixtures and PCA may be used to interpret any multivariate data, such as, for example, water quality, based on the measurements of dissolved oxygen, temperature, salinity, turbidity, suspended solids, nitrogen, and phosphorus [69].

The profiles of hydrocarbons, PCBs, organochlorine pesticides, and sterols in sediments from the northwestern Mediterranean Sea were studied by PCA, hierarchical cluster analysis, and positive matrix factorization [70]. Three sources could be distinguished: anthropogenic, consisting mostly of PAHs; rivers, containing mostly *n*-alkanes, pesticides, and sterols; and an "unspecified background" source, containing just some *n*-alkanes.

Principal Component Analysis (PCA) is being used increasingly to interpret multivariate data, such as concentrations of metals in sediments. PCA could clearly separate metal patterns in clean, less-clean, and highly contaminated sediments and showed correlations between Co and Mn, Zn and Pb, and a relatively weak relationship between Cu and Cd [71].

The above examples of PCA applications deal with a two-way data analysis (subjects and variables). A three-way PCA (subjects, variables, and conditions) has shown that trace metal concentrations can be used to classify healthy and diseased blue crabs [72].

Nonlinear mapping has been used to study PAHs in the Bay of Biscay [73]. This technique projects the multidimensional data on a plane in such a way that the similarity of samples (distances between points) are preserved as much as possible. This is in contrast to PCA in which orthogonal projections are made on planes oriented in the direction of the largest variations in the original data. PCA projections are easier to visualize and use in subsequent calculations than nonlinear maps.

6
Determination of Trends and Effects

Two important questions in marine pollution studies are: (1) are regulations of uses of chemicals and discharges of wastes reflected in improved quality of the marine environment?; (2) has pollution affected the health of the oceans as judged by the abundance and health of marine fauna and flora?

The determination of temporal trends in concentrations of pollutants in water, sediments, or biota, usually over a period of years, is required to answer the first question. The US Mussel Watch data show a significant linear decrease for chlordane, dieldrin, DDT, and PCB at most sites along the US coast, but cadmium and copper decrease at only a few sites. No trend was detectable for PAHs and zinc. An increasing trend was detected for the latter at one site [74]. Similarly, a decrease in the concentration of PCBs and DDT was observed in cod and herring from the Northern Baltic [75]. Herring is a species particularly difficult to use for temporal trend studies since its concentration of lipids tends to fluctuate quite widely from year to year. The concentrations of contaminants are affected by many factors in addition to the input, and the determination of temporal trend may require many analyses not only of the pollutants, but also of additional variables. For example, the addition of temperature and rainfall improved the time-trend model for concentration of chlordane, PCB, and dieldrin, but had little effect on the concentration of most heavy metals [74].

Some general patterns in time trends of organochlorine compounds have appeared [76]. After the cessation of input, the relatively high levels near the source decrease exponentially. The change is less pronounced at a distance from the source and the decrease may resemble a parabola. Low concentrations at distant sites may level off. For aquatic biota these patterns are exhibited by short-lived fish for the first pattern, and by long-lived fish, marine mammals, and invertebrates for the other two trend patterns.

Statistical consideration for the design of monitoring programs in relation to monitoring objectives have been described and a table of the probabilities of detecting a linear trend as a function of the length of the monitoring period and the signal/noise ratio (regression coefficient/standard deviation of regression) have been published [77]. For example, in a five-year program there is a 90% chance of detecting a trend with signal/noise ratio of 1.58.

Biological effects monitoring is an essential part of marine pollution studies which, in comparison with chemical analyses, has been relatively neglected. Biological effects monitoring is a very difficult task because of the complexities of the marine environment. Because of cost, the monitoring may be carried out only on a limited number of species and there is no guarantee that important species will be selected. The environment is simultaneously exposed to many factors which may interact and, consequently, the isolation of the effects of a particular factor and predictions of impacts, except for the most obvious ones, may be practically impossible [8].

A "Joint Assessment and Monitoring Program" (JAMP) has been agreed internationally for the northeastern Atlantic, the North Sea, the Celtic Sea and coastal zones of Western Europe [78]. The program will include contaminant-specific techniques and effects monitoring when specific contaminants are unknown. The contaminant-specific measurements include those of PAH metabolites in bile, cytochrome P4501 A (EROD) in liver and liver pathology, mercury, cadmium, and lead metallothionein in liver, ALA-D in blood, antioxidant defenses in liver, and liver pathology, and for TBT imposex/intersex in gastropods and shell thickening in *Crassostrea*. The suggested parameters for general effects monitoring include cytochrome P4501 A (EROD), lysosomal stability, liver pathology, reproductive success in the viviparous blenny, external fish diseases, benthos community structure, and the occurrence of liver nodules. In addition, when an impact is suspected, water, sediment, and pore water, caged or sedentary animal bioassays, and fish reproductive success are recommended.

To answer the second question on whether pollution has affected the health of the ecosystem is even more difficult, since one has to deal with natural variations and interactions of faunal and floral populations. In the case of populations exploited commercially, there is the need to separate natural from exploitation mortality. This problem has been discussed in detail [79]. There are examples of localized effects, but no evidence of major detrimental changes that could be directly related to pollution. Thus, for example, a number of pathological conditions in fish, such as hepatic neoplasms and fin erosion, have been associated with sites containing high sediment concentrations of polynuclear aromatic hydrocarbons [80–82]. On the other hand, the reasons for the world-wide decline of marine fish landings are still uncertain.

7 A Brief Overview of Classes of Substances Related to Marine Pollution

7.1 Heavy Metals

A voluminous literature deals with heavy metals, particularly mercury, cadmium, lead, copper, and zinc in the marine environment [42]. The concentration of mercury, cadmium, and lead in open ocean sea-water are about 0.3–10 ng/l, 0.2–20 ng/l, and 10–60 ng/l, respectively. The concentrations in coastal water may be about 3–5 times higher. There are limited data on the concentrations in deep-sea surface sediments, except for lead, for which a recent anthropogenic input has been documented [42].

Interlaboratory CVs for heavy metals in sediments were, on average, 16%. Only for mercury was the CV 50%. On average, 80% of the results were within two standard deviations of the accepted mean. The CV for organic carbon and for dry weight of biota were 16% and 6%, respectively. The results for heavy metals in biota were much less satisfactory, with 46% of the laboratories reporting concentrations of As, Cd, Cu, and Hg within two standard deviations and only 8% reporting such values for Ni and Cr [83].

The problem of the speciation of metals in sediments continues to attract attention, both from the point of measurement techniques and from the point of bioavailability and, consequently, effects such as toxicity. Some workers prefer to measure total metals after full digestion of sediments with hydrofluoric acid, others favor sequential extractions [84]. The concentrations of heavy metals in coastal sediments are remarkably similar world-wide [85]: Cu 30 μg/g, Ni 30 μg/g, Pb 40 μg/g, Zn 120 μg/g, Ag 1 μg/g, Cd 1 μg/g, Hg 0.5 μg/g dry sediment containing at least 20% of silt and clay, respectively. "High" levels are about 3–5 times higher. Concentrations of metals in settling particulate matter may respond faster to changing input than concentrations in surficial sediments. Thus for example the concentration of lead in settling particulates in the Bothnian Sea has decreased 50%, which is in agreement with aerial fallout estimates. On the other hand, the load of cadmium might have increased by 70% [71].

Relatively little is known about antimony in the marine environment. It may require a closer look if an input from mining or flame retardant production was anticipated. Water in Osaka Bay contained antimony at 0.15 µg/l, with one sample as high as 0.8 µg/l. Higher concentrations were reported for fresh and rain water [86]. Since, historically, many major pollution problems started in Japan (Minamata – mercury, itai-itai – cadmium, yusho – PCBs), these findings may well be an advance notice to initiate a survey of antimony in the marine environment.

Algae, mollusks, and invertebrates tend to concentrate heavy metals to a larger extent than fish. Fish accumulate mainly mercury in the form of methyl mercury to quite high levels, particularly in large, long-lived, and slow growing species such as large tunas and sharks. Documented relationships between the concentration of heavy metals in tissues and response of marine fauna are rare. The majority of studies have concentrated on human safety of "fish as food" instead. The form of storage of heavy metals, particularly in mollusks and invertebrates, is another interesting question that has not received the attention it deserves.

With the exception of tributyl tin, no deleterious effects could be assigned to individual heavy metals in the United Kingdom estuaries [87].

7.2
Other Inorganics

Arsenic concentration in sea-water is about 2 µg/l and arsenic is present primarily as arsenate. Sediments may contain arsenic in concentrations up to 40 µg/g dry weight, again as arsenate. Algae take up arsenic from sea-water and metabolize it via arsenite to a number of organoarsenic compounds. Arsenic is widely encountered in marine biota, particularly in the form of arsenolipids, mainly arsenobetaine. The biochemical reasons behind this are not known and the topic is of relatively low priority because of low human toxicity of these compounds [88, 89].

Selenium may be a problem in the freshwater environment; 5 µg/l is the criterion for protection of aquatic life at chronic exposure and modifications of this criterion have been proposed recently [90]. A similar situation apparently does not exist in the marine environment. Selenium is a relatively abundant trace element in the marine environment and seems to counteract the toxic effects of mercury.

Copper, chromium, and arsenic may be leached from wooden structures such as wharves preserved by chromated copper arsenate (CCA). CCA is toxic to aquatic biota in laboratory tests. In the field the effects are limited to a few meters in the vicinity of the preserved structures [91].

7.3
Nutrients

Nutrients have traditionally been a part of marine chemistry but, with increasing concentration of nutrients in coastal waters [92], signs of eutrophication, and an increasing frequency of harmful algal blooms that may produce toxins affecting human health [93], nutrients are now more often incorporated into pollution studies. Some of the research problems include the predictability of blooms and species sequences from nutrient concentrations, the detection of toxins, the characterization of general "organic" enrichment of marine waters, and the interaction of algal excretion products with contaminants.

For phosphates at "high" and "low" concentrations in sea-water (1.33 μmol/l and 0.069 μmol/l, respectively), the coefficients of interlaboratory variation were 8% and 64%. For nitrate+nitrite the CVs were 5% and 18% for concentrations of 17.35 μmol/l and 1.45 μmol/l, respectively. CVs of 20% and 57% were obtained for ammonia at 3.33 μmol/l and 0.38 μmol/l, respectively [94].

7.4 Petroleum Hydrocarbons

Hydrocarbons are the most frequent and visible chemicals spilled in the oceans. Of the 2175 slicks detected from 1992 to 1995 in the North Sea, 826 were caused by hydrocarbons, 71 by vegetable oils, 4 by chemicals, 15 by other causes, and 1223 were of unknown origin [95]. Oils predominate in accidents with environmental consequences [96]. A vast literature exists on the measurement, fate, and effects of petroleum hydrocarbons in the marine environment [97]. Large oil spills have spectacular, catastrophic, but not lasting effects on the environment and the main topics of interest are prevention and cleanup. More attention is needed for less noticeable, chronic petroleum hydrocarbon releases. In addition, the use of hydro-treated, low aromatics petroleum products has increased over the years. These products contain relatively high levels of cycloparaffins such as the decalines and tetralines, whose fate and effects on the environment are largely unknown. Analytical methods should be selected according to the objectives of the analyses. The objectives may be the identification of source, measurement of the contamination of the environment for damage assessment, the determination of safety of seafood, or the impact of the oil on marine fauna or flora [98]. Many commercial products such as pesticide formulations, hydraulic or transmission fluids, lubricating and cutting oils have the appearance of "oil" and, unless the source of a spill is known with certainty, oil spills should be investigated by a competent chemical analytical laboratory, able to detect unanticipated chemicals other than hydrocarbons, to determine the identity of the "oil".

Large amounts of data on hydrocarbons in water, sediments, and biota are available for the Baltic, the North Sea, the Atlantic, and the eastern Pacific, and data from southeast Asia are also starting to appear in the literature [99].

Offshore oil production requires many chemicals for well drilling, for the treatment of the produced oil, for the treatment of gas, and for the stimulation and workover of the wells [100, 101]. Altogether, 25 classes of chemicals are used in offshore oil production [102]. The chemicals are undergoing increasingly detailed testing. As in other Quantitative Structure-Activity Relationships (QSAR), the octanol/water partition coefficient (K_{ow}) is a crucial factor determining the environmental behavior and toxicity of the chemicals [103]. Toxicity testing is an essential part of hazard identification and laboratories conduct intercalibrations [104].

Many of these "chemicals" are complex mixtures of compounds and range from drilling muds through corrosion inhibitors, coagulants, and surfactants to biocides. Drilling chemicals are used in largest quantities and consist mainly of inorganics such as bentonite and barite. These are the main components of "drilling muds" that may be either water- or oil-based. The former usually contain lignosulfonates as dispersants and glutaraldehyde as a biocide. The oil in oil-based muds is most frequently a hydrogenated petroleum product, which is very low in aromatics, but rich in various cycloparaffins. Synthetic mud base fluids include $C_{11}-C_{16}$ *n*-alkanes, a hydrogenated decene dimer, a series of isomeric C_{16} and C_{18} alkenes, a series of C_{14} and C_{16} monoenes, C_8-C_{16} carboxylic acids esterified with 2-ethylhexan-1-ol, and di(2-ethyl-hexan-1-ol) isobutyl acetal [105].

Emulsifiers in oil-based drilling muds are usually salts of fatty acids or resin acids. Some drilling muds may also contain defoamers, such as tributyl phosphate. About one half of the water-based drilling mud used is released into the sea and its effect is mostly physical. The discharge of oil-based drilling muds is generally restricted and even the discharge of oily cuttings may be prohibited.

Production chemicals are used in relatively much smaller quantities, but significant proportions are discharged into the environment. The chemicals include biocides, scale and corrosion inhibitors, flocculants, surfactants, and paraffin inhibitors. The commonly used biocides include glutaraldehyde, quaternary amine salts, and, to a much smaller extent, some thiocarbamates, isothiazolin, and hypochlorite. Other biocides include Bioban P-1487 (80% 4-(2-nitrobutyl) morpholine, 10% 4,4′-(2-ethyl-2-nitrotrimethylene) dimorpholine, 4% 1-nitropropane), and Vantocil IB, an aqueous solution containing 20% polymeric biguanidine hydrochloride [104].

The corrosion inhibitors are usually heterocyclic amines. The flocculants are polyamines, surfactants are alkyl aryl sulfonates and ethoxylated alkyl phenols, and the paraffin inhibitors are usually alkyl or aryl polyethers.

7.5 Polynuclear Aromatic Hydrocarbons (PAHs)

PAHs in the marine environment are receiving considerable attention, partly because of their carcinogenicity and partly because they are comparatively easily detectable. PAHs are always released into the environment in mixtures that may be characteristic for different sources [106, 107]. The most important point sources for the marine environment are aluminum and manganese smelters [108], wood-preserving plants using creosote, petroleum refineries, and, to some extent, storm sewers. General combustion processes contribute PAHs by long-range fallout. Between 50% and 75% of PAHs entering the sediment of the western Mediterranean Sea annually originates in atmospheric fallout [109]. In sediments the concentration of PAHs is strongly correlated with that of soot carbon. This indicates that PAHs are adsorbed on soot and much less bioavailable than one would expect from adsorption on organic matter in general [110].

The concentrations of PAHs with more than two aromatic rings in sea-water are very low. On the other hand the concentrations of all PAHs in sediments may be very high.

Marine invertebrates, particularly mollusks, accumulate PAHs and tend to retain the profile of the source of PAHs [111]. PAHs in mussels transplanted from a clean area reached significantly higher levels three years after an oil spill [112]. On the other hand, PAHs are metabolized in fish by monooxygenases and epoxide hydrolases to various hydroxylated derivatives that are usually excreted as glucuronides, but may also be bound covalently to various macromolecules such as DNA, RNA, or proteins [113, 114]. The presence of PAH glucuronides in the bile and the induction of the enzymes are indicators of exposure. The acute toxicity of PAHs to fish is relatively low and decreases with increasing molecular weight. On the other hand, chronic toxicity may manifest itself by various sublethal responses such as immune suppression, DNA and RNA adduct formation, eye lens cataracts, and liver lesions, at concentrations in the µg/l range. The toxicity of PAHs may be enhanced by sunlight. Effects ranges for total PAHs, derived by "hockey stick" regression from liver lesions in English sole [115], 230–2800 ng/g dry sediment, are lower than effect range-low (ERL) [38], 4000 ng/g dry sediment.

Creosote, used for the preservation of wooden structures, is the most common source of PAHs in the coastal marine environment. "Weathered" creosote (creosote fractions containing mostly higher molecular weight PAHs) in sediments has a higher chronic toxicity for fish than the lower molecular weight fractions, which, in turn, are much more acutely toxic [116].

Interlaboratory coefficients of variation for PAHs in standard solutions and in sediment extracts were 18% and 24%, respectively [117].

Concentrations of total PAHs in bivalves from the Caribbean and Central and South America ranged from 28.4 ng/g to 13,800 ng/g dry weight [31]. As expected, the highest concentrations were observed near ports or urban centers. Only three PAH profiles, one each from Chile, Brazil, and Argentina, are shown in the report. The latter is conspicuous by relatively low concentrations of naphthalene and alkyl naphthalenes, indicating primarily a combustion source.

7.6 Synthetic Organic Chemicals

Many synthetic organic chemicals are produced, shipped, and used commercially and may reach the marine environment. Pure chemicals would most likely be released by shipping accidents and, in small quantities, by leaching from technical materials or disposal of products such as "light sticks". The majority of synthetic organic chemicals reach the marine environment as trace components of complex chemical "soups" either in direct releases in water or as short- or long-range fallout from the atmosphere. Organic chemicals are generally less soluble in sea-water than in freshwater. Consequently, in sea-water they appear more "hydrophobic" and the partitioning equilibria are shifted towards air, organic carbon, and lipids; these factors must be taken into consideration in studies of the behavior of organics in sea-water [118].

The selection of chemicals for study and the priority ranking is the subject of a large number of publications. The factors to consider in the selection are persistence, toxicity, amount likely to be released to the environment, alternative chemicals, and the possibility of corrective action. The precautionary principle is applicable here. It is easier to prevent a problem by not using a chemical than to treat the effluent, and it is impossible to remove a released chemical from the environment.

Most of the chemicals of highest environmental concern have very low solubility in water and, as a consequence, may partition to the atmosphere or be associated with particulate matter. The former may lead to long-range transport, particularly to colder regions of the globe. The latter may at least temporarily limit their bioavailability from the water column [119] and affects their transport in the environment towards depositional sinks such as certain zones in estuaries [120].

The strength of the carbon-fluorine bond, the isostericity of fluorine with the hydroxyl group, and the strong electronegativity of fluorine confer special properties on organofluorine compounds. For example, many enzyme inhibitors contain fluorine, and perfluorinated alkyl chains are both hydrophobic and lipophobic. The number of organofluorine pesticides, mostly herbicides, has more than doubled over the past 15 years and the majority of them contain the very stable trifluoromethyl group [18]. In the analytical laboratory, fluorine-containing compounds may escape attention since, in contrast to chlorine and bromine, fluorine lacks a characteristic isotopic "signature".

7.6.1 *Pesticides*

Organochlorine pesticides of the DDT and BHC groups, dieldrin, and some components of chlordane are frequently present in many sediments [42] and are almost always found in marine biota [42]. In sediments, the concentrations of DDT and metabolites and BHC range from ng/g to µg/g. Toxaphene is reported less frequently, but it often may not be recognized when present in low concentration or modified by biodegradation. The levels of many of these pesticides are still quite high, particularly in top predators. This is understandable because of the considerable persistence of these compounds and the fact that some are still used, particularly in the developing countries. There is also evidence that these compounds, released in the tropics, are then transported aerially to colder regions [121], with particularly the arctic ecosystems acting as a sink [122].

Concentrations of organochlorine pesticides in bivalves from the Caribbean and Central and South America have been reported under the IMW project [31]. The concentrations of DDT and metabolites were comparable to those found in bivalves from the United States coasts, with the majority of samples containing total DDT at less than 50 ng/g dry weight, and a few samples with total DDT at 100–200 ng/g dry weight. Concentrations of lindane were mostly < 5 ng/g.

Chlorinated bornanes, PCBs, DDT and metabolites, components of chlordane, hexachlorocyclohexanes, and chlorinated benzenes were measured in the blubber of harbor porpoises from several locations in the western Atlantic [123]. Chlorinated bornanes, originating from toxaphene, were predominant in porpoises from the most northern locations – Newfoundland and the Gulf of St Lawrence. It is surprising that only 46% of DDT and metabolites was present as DDE. Of the chlordane components, *trans*-nonachlor was the most predominant. Chlorobenzenes consisted of penta- and hexachlorobenzene; their concentrations were higher in males and increased in the northerly direction. The α-isomer was the main hexachlorocyclohexane present. Mature males had considerably higher concentrations of organochlorine compounds than mature females. The concentrations have declined since the 1970s by factors of 4 and 17 for PCBs and DDTs, respectively in the Bay of Fundy. In other areas the decreases since the 1980s were by factors of 2–4.

In contrast to organochlorine pesticides, the newer organophosphate, carbamate, and pyrethroid pesticides are seldom detected, partly because they are less persistent, partly because low concentrations are much more difficult to detect than those of most organochlorine compounds.

A pesticide with the most pronounced effect on the marine environment is tributyl tin. It is extremely toxic to some groups of marine fauna, such as whelks and shrimp [124]. Many countries took action to limit the release of tributyl tin (TBT) into the marine environment by banning its use on boats < 25 m in length, but the initially rapid decline of TBT concentration in sea-water has leveled off and the concentration remains above the environmental quality objective of 2 ng/l (10 ng/l in The Netherlands). It may be that still more stringent actions are required. The main problem is that there is thus far no good replacement. One potential substitute is Irgarol 1051(2-methylthio-4-*tert*-butylamino-6-cyclopropyl-*S*-triazine). Unfortunately, it is biodegraded only slowly, if at all, and there are already reports of its presence in the marine environment. The green alga *Enteromorpha intestinalis* is very sensitive to Irgarol 1051 and could possibly be used as an indicator species [125].

The estimated annual input of TBT into the North Sea is 68 tonnes, primarily from commercial vessels. Approximately 20% of the world's commercial shipping passes through the North Sea. A model of TBT inputs and fluxes indicates that areas of highest TBT impact are the east coast of the UK, the Skagerrak, and the English Channel [126].

The herbicide mecoprop (MCPP, 2-[4-chloro-2-methylphenoxy]propionic acid), as well as a related compound, clofibric acid (CA, 2-[4-chlorophenoxy]-2-methylpropionic acid), which in the form of ethyl ester is used in human medicine to regulate blood lipids, were detected in the North Sea water in ng/l concentrations [127]. A series of triazine herbicides including simazine, atrazine, propazine, terbutylazine, and prometryn, and their metabolites desethylatrazine and desethylterbutylazine, as well as the phenylurea herbicide linuron, were detected in the sediments of the German Wadden Sea [128].

7.6.2
Other Industrial Organics

This group contains many organochlorine and organobromine compounds such as chlorinated benzenes, PCBs, polychlorinated terphenyls, polybrominated biphenyls (PBBs), chlorinated paraffins, chlorinated and brominated diphenyl ethers, chlorinated and brominated cycloparaffins. Some of the compounds, for example hexachlorobenzene and mirex, may also be used as pesticides.

The production of many of these compounds, such as PCBs, PBBs, and some chlorinated paraffins, has been discontinued, but the compounds will remain in the environment for a long time because of their extreme persistence. Considerable effort is devoted to the measurement of these compounds in the environment.

There is a great deal of uncertainty in the measurement of the concentration of PCBs as well as of other highly hydrophobic chemicals in sea-water because of their extremely low solubility, strong tendency to adsorb on particles (including the sampling equipment), and possibility of contamination during sample processing [129]. These problems are not as severe in sediments and biota, but, in all cases, an additional uncertainty is introduced by the method of quantitation. The early measurements used commercial PCB formulations as standards (for example Aroclor 1254). Later data are usually given in terms of individual chlorobiphenyls or their sums. Sometimes, only the non-*ortho*-, and mono-*ortho*- chlorobiphenyls are reported, because of their TCDD-like toxicity. The latter does not mean that the di-*ortho*- substituted chlorobiphenyls are not toxic, they just lack this type of toxicity and appear to be neurotoxic instead [130].

Fowler [42] summarized data on PCBs published up to the late 1980s. The concentration of PCBs in sea-water is in the pg/l to ng/l range and in sediments starts at ng/g of dry sediment and may reach the µg/g range in heavily contaminated areas. Since sea-water and sediments are very large compartments, a small uncertainty in PCB concentration has a pronounced effect on estimates of PCB budgets. Early estimates of the amount of PCBs in sea-water ranged from 6000 to 66,000 tonnes, in marine sediments 660 to 2700 tonnes, and in marine biota, 30 tonnes [131]. More recently, a budget for the western Mediterranean was estimated. The levels of PCBs in the western Mediterranean have declined in the last 15 years, but remain relatively high in some coastal areas. The amount of PCBs present is about 140–420 tonnes; inputs consist primarily of aerial deposition (12 tonnes/year), as compared to freshwater input of 1 tonnes/year [132]. The Baltic may contain about 100 tonnes of PCBs in the water column, and 26,000 tonnes in the sediments [133].

Interlaboratory coefficients of variation of chlorobiphenyls and organochlorine pesticides in sediments are about 40%, those in cod liver oil, about 26% [134].

In the North Sea, concentrations of PCBs have recently been expressed as the sum of the congeners 28, 52, 101, 118, 138, 153, and 180. In these units, the concentrations ranged from 16 pg/l to 30 pg/l and were about four times those in the eastern Atlantic [29]. The average concentration in sediments, expressed in terms of the congener 153, was 0.57 µg/kg.

Levels of PCBs in bivalves from the Caribbean, Central America, and South America, reported by the IMW Project [31], appeared similar to those found in bivalves from the US Gulf of Mexico coast, with the chlorobiphenyl 138 concentrations mostly below 25 ng/g dry weight. A more detailed evaluation of PCB profiles in the IMW samples is not available as yet.

In harbor porpoises from the west Atlantic, PCBs were the second most predominant contaminants in most locations and ten chlorobiphenyls (IUPAC number (and chlorine positions)) 52 (25–25, possibly with 235–2), 95 (236–25), 101 (245–25), 151 (2356–25), 149 (236–245, possibly with 2356–24), 118 (245–34), 153 (245–245, possibly with 234–236), 138 (234–245, possibly with 2356–34), 187 (2356–245), and 180 (2345–245 possibly with 2356–345)) accounted for over 50% of the total concentration on a weight basis. The chlorobiphenyls 153 and 138 amounted to 23–30% of the total. The PCB profiles were similar in all locations [123].

Chlorobiphenyl profiles did not reflect trophic status of biota at the Midway Atoll (North Pacific) and showed considerable differences even for species of the same trophic level (Figs. 3 and 4, from the data of Hope et al. [135]). The chlorobiphenyl profiles of sea urchins, snails, and fish are quite similar and dif-

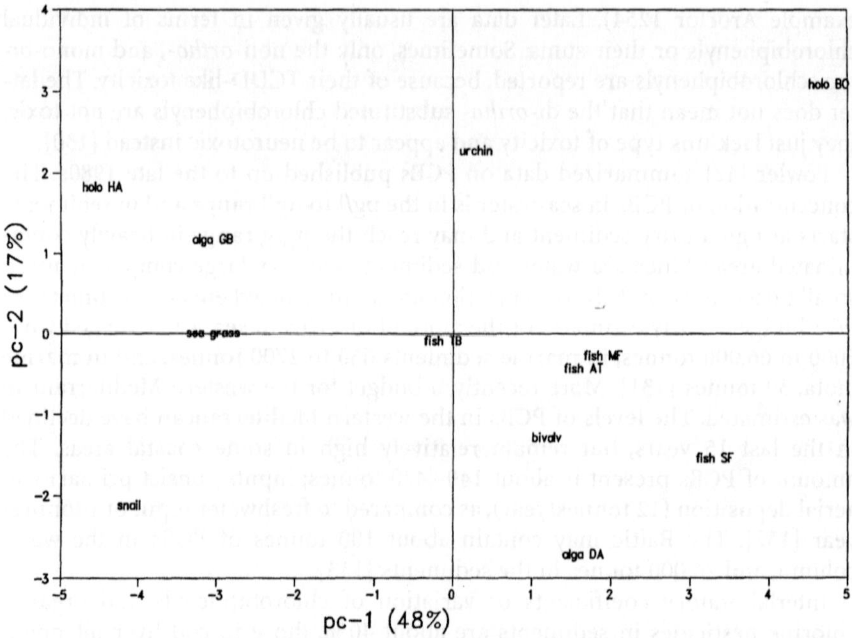

Fig. 3. Chlorinated biphenyls in marine species from Midway Atoll, projected on the plane of the first two principal components. The portions of original variance accounted for are indicated on the axes. Calculated from the data of Hope et al. (1997) [135]. Species are indicated in the figure: algae *Dictyota acutiloba* and *Giffordia breviarticulata*, sea grass *Halophila ovalis*, bivalve molusk *Chama iostoma*, sea urchin *Echinometra mathaei*, snail *Nerita picea*, holothurians *Bodadschia obesus* and *Holothuria atra*, and fish *Acanthurus triostegus*, *Mulloidichthys flavolineatus*, *Stegastes fasciolatus*, and *Thalassoma ballieui*

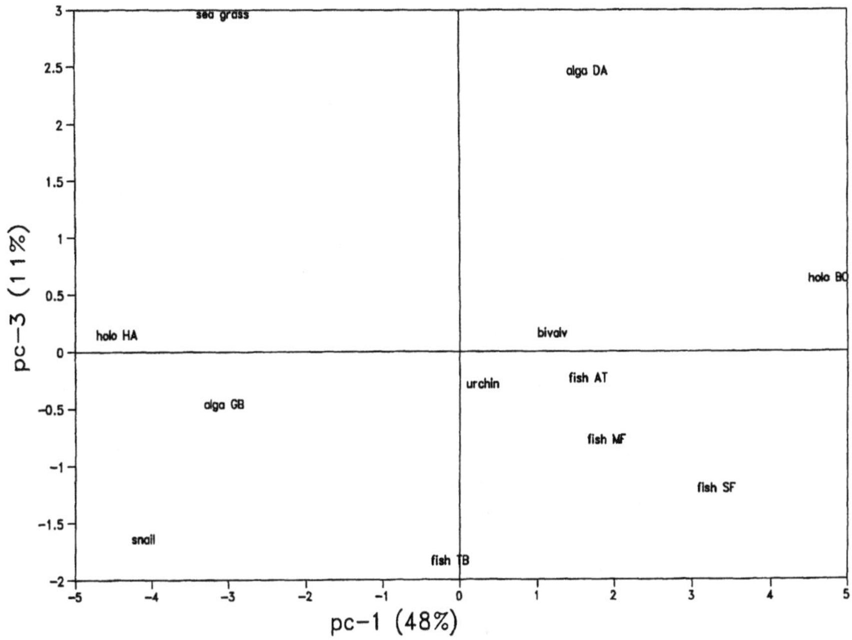

Fig. 4. Chlorinated biphenyls in marine species from Midway Atoll, projected on the plane of the principal components 1 and 3. The portions of original variance accounted for are indicated on the axes. Calculated from the data of Hope et al. (1997) [135]. Species are indicated in the figure: algae *Dictyota acutiloba* and *Giffordia breviarticulata*, sea grass *Halophila ovalis*, bivalve molusk *Chama iostoma*, sea urchin *Echinometra mathaei*, snail *Nerita picea*, holothurians *Bodadschia obesus* and *Holothuria atra*, and fish *Acanthurus triostegus*, *Mulloidichthys flavolineatus*, *Stegastes fasciolatus*, and *Thalassoma ballieui*

fer from those of algae, sea grass, and holothurians. Interestingly, the chlorobiphenyl profiles of the two holothurian species are very different. It may be that analytical bias contributes to some of the differences. For example, the concentration of the chlorobiphenyl 209 (decachlorobiphenyl) in the alga appears to be unusually high; this chlorobiphenyl was not used in the PCA of the data.

The continuing preoccupation with PCBs is to some extent an exercise in futility since nothing can be done to alleviate the problem. More effort should be made toward hazard identification, risk assessment, and possible replacements for organohalogen compounds, such as the halogenated diphenyl ethers and chlorinated and brominated cycloparaffins, that are still being used and released into the environment and for which there are many data gaps.

Polychlorinated naphthalenes (PCNs) have recently received increased attention [136]. In contrast with PCBs, PCNs have not been in widespread use for many years and their levels in the environment are likely lower than those of the PCBs. Sources of CNs are the commercial products, but CNs are also present as impurities in commercial PCBs and may be formed during incineration of municipal waste and in electrochemical processes based on graphite electrodes. CNs at 7.6 ng/g dry weight were found in one sediment sample from the Baltic.

Their levels in Baltic herring and cod ranged from 9 ng/g to 26 ng/g lipid [137]. The sediment sample contained a large number of chlorinated naphthalenes, whereas only a few were present in the fish.

Polychlorinated diphenyl ethers (PCDE) were found in the ng/g range in cod liver oils [138]. The PCDE profiles were compared with those of PCDEs in chlorinated phenols, wood preserving formulations, and fly ash. The primary source of PCDEs are chlorophenols but sources other than those investigated appear to be involved as well. The majority of PCDEs found in cod liver oil are mono- or di-*ortho*- substituted and also have chlorines in the *meta* and *para* positions.

Polychlorinated dibenzodioxins (PCDD) and dibenzofurans (PCDF) also belong under this heading, except that they are not produced intentionally. They are formed at elevated temperatures form various organic compounds in the presence of chlorine compounds or even the chloride ion. A few of the compounds, particularly 2,3,7,8-tetrachlorodibenzodioxin (TCDD) and 2,3,7,8-tetrachlorodibenzofuran (TCDF) and some penta- and hexa-chloro dibenzodioxins and dibenzofurans containing chlorine in these and some additional positions, are extremely toxic to some species of marine mammals. PCDD/Fs occur in sea-water in ultratrace concentrations of the order of fg/l, in sediments, and in biota in concentrations of the order of pg/g. Accordingly, the analyses are very costly and, again, the best course of action is to limit the formation of these compounds in the first place. Their mechanism of toxicity is similar to that of non-*ortho*-chlorinated biphenyls. The former are about 1000 times less toxic, but may be present in 1000 times higher concentrations and be toxicologically more significant than PCDDs and PCDFs. Interestingly, polyaromatic compound fractions of sediments from the west coast, and even more from the east coast, of Sweden displayed more TCDD-like activity (induction of the enzyme 7-ethoxyrufin-*O*-deethylase (EROD)) than the PCB-PCDD-PCDF fractions. The EROD-active compounds remain to be identified [139].

Tetra- and penta-bromodiphenyl ethers and their methoxy derivatives have been detected in herring, salmon, seals, and commercial fish oils from the Baltic [140] in the tens to hundreds of ng/g lipid range, with levels increasing up the food chain. The samples also contained polychlorinated diphenyl ethers ranging from hexa- to decachloro and methoxy derivatives of octa- and nonachloro diphenyl ethers.

The compounds tris(4-chlorophenyl)methanol and, to a lesser extent, tris(4-chlorophenyl)methane have been detected in common seal, dolphin, white-beaked dolphin, and harbor porpoise blubber, mussels, cod liver, and sediments from the North Sea and from the Wadden Sea [141]. The source and toxicity of these compounds are not known. The compounds may be related to triphenylmethane dyes or may be by-products in the production of DDT.

Chlorinated fatty acids are present in marine fauna in high concentrations (tens to hundreds of µg/g lipid) and brominated unusual fatty acids have been found in some sponges. Some of the chlorinated fatty acids may originate directly from pulp mill effluents or, hypothetically, be formed by metabolism of chlorinated paraffins [142].

2,4,6-Tribromophenol and, to a lesser extent, 2,6-dibromophenol, found widely in marine fauna in the ng/g range [143], are of natural origin.

Another important group of industrial chemicals are phthalates. These are used mainly as additives to plastics, but may also have other uses such as solvents for pesticides. They are much less toxic and persistent than many of the organohalogen compounds, but are used in larger quantities. Somewhat related to phthalates are other esters such as adipates and citrates, but they are biodegraded more rapidly because of the absence of the aromatic ring of phthalates.

Relatively little is known about industrial organophosphates, such as triphenyl phosphate, trioctyl phosphate, and tris(chloropropyl) phosphate. These compounds are used as plasticizers and flame retardants and are likely to reach the marine environment. Six organophosphates were detected in the water of Osaka Bay in concentrations ranging from 0.1 μg/l to 1.3 μg/l. The compounds tris(2-chloroethyl)phosphate and tributyl phosphate were the most abundant [144].

It is practically impossible to identify all chemicals in a sample and yet such procedures are needed to detect the presence of unanticipated chemicals and to signal new or potential problems. Extracts of water, sediment, and biota have been extensively fractionated and analyzed by multidetector gas chromatography [145]. Compounds detected in addition to the usual hydrocarbons, chlorinated hydrocarbons, and organochlorine pesticides included chlorinated isocyanates and anilines, cyclohexanol, cyclohexanone, aromatic ketones and acids, thiophenes, nonyl- and cumyl-phenols, carbazoles, quinolines, indoles and tolyltriazoles, and the pesticides permethrin, triclosan, and bromacil.

7.6.3
Plastics

Discarded plastic objects as well as plastic globules destined for the production of plastics are found in the marine environment [146]. Not only is their presence unsightly, but they may also be harmful to marine fauna by physical action, such as interference with feeding, with digestion, or "ghost" fishing by abandoned or lost netting. Low-molecular-weight additives may be leached from the plastics, or be ingested with the plastics, by aquatic biota. For example, many expanded polystyrene globules contain the flame retardant hexabromocyclododecane [147]. Research can play only a minimal role in preventing pollution by plastics and other debris. The best course of action is to document the situation and exert pressure on society not to litter [148].

7.6.4
Surfactants

Concentrations of alkyl benzenes [149] and trialkyl amines in marine sediments have been reported and used as tracers of domestic sewage discharges. Little else is known about the fate of surfactants in the marine environment. Except for very localized areas in the vicinity of municipal and some industrial discharges, the concentrations appear to be well below those associated with observable effects.

The most persistent nonionic surfactants are based on alkylphenol ethoxylates. These ethoxylates are only moderately accumulated by marine fauna. The bioconcentration factors in common mussels (*Mytilus edulis*) decreased from 350 to 50 from nonylphenol to nonylphenol tri-ethoxylate [150].

Typical linear alkylbenzene sulfonate (LAS) concentrations in Dutch coastal waters decrease from about 5 µg/l in estuaries to 1 µg/l in close inshore samples, and below the detection limit of 0.5 µg/l in offshore (>10 km) samples [151]. LAS are degraded by oxidation and shortening of the alkyl chain by one or two carbon atoms. The products of this process are sulfophenylcarboxylic acids [152]. They appear somewhat less toxic than the parent compounds, but few details are available.

Foams quite often visible in some estuaries, particularly after periods of heavy freshwater runoffs, are usually caused by humic substances.

7.7 Complex Effluents

Many sources consist of ill-defined mixtures often characterized only by nonspecific parameters such as chemical or biological oxygen demand, total dissolved or suspended solids, etc. Studies of such effluents usually include toxicity determination of the whole effluent and, if possible, the identification of a component or a property that can be used to study the dispersion of the effluent in receiving waters.

7.7.1 *Sewage*

Many municipalities are still discharging raw sewage. Typically, raw sewage contains 99% water and 1% total solids. Of the latter, 70% consists of organic materials, mostly proteins, some carbohydrates and fats, and 30% is inorganic [153]. The trend is towards the implementation of primary and secondary treatment, which decrease the BOD by 30% and 90% and suspended solids by 70% and 90%, respectively. A major problem, particularly in larger municipalities, is the collection of sewage from numerous outfalls into the treatment plant. The main effects of sewage discharge are oxygen depletion and eutrophication.

It is useful to be familiar with the treatment process, to monitor the changes that occur after the introduction of the treatment, to characterize chemically and microbiologically the treated effluent, and to determine its dispersion in the receiving waters. It appears that, at least in the UK, the impact of sewage discharge on the marine environment is relatively minor and the main factor responsible for this is high initial dilution [154]. Without sufficient dilution, eutrophication is the most likely consequence, as demonstrated for example by a case study [155].

Trialkyl amines [156] and linear alkyl benzene sulfonates are good tracers of municipal effluents in the marine environment.

Increased organic loading is usually limited to the vicinities of the aquaculture sites and to local depositional areas [164, 165]. Nutrients and some other feed additives may cause algal blooms on a wider scale. Attempts are underway to reduce the input of phosphorus and nitrogen. The Nordic aquaculture has already achieved input into the Baltic which is negligible in comparison with other sources, but still significant in local areas [166]. Further technological improvements such as extruded rather than pelleted feed, with higher lipid and lower nitrogen content, automated feeding, and oxygen injection are underway [161].

Little is known about the long term effects of the biocides. The assortment of these compounds is changing fairly rapidly, with research trying to catch up. For example, Atlantic salmon were treated for sea lice infection by DDVP (dichlorvos, 2,2-dichloroethenyl dimethyl phosphate), trichlorfon (dimethyl (2,2,2-trichloro-1-hydroxyethyl)phosphonate), hydrogen peroxide, ivermectin, azamethiphos (*S*-[(6-chloro-2-oxooxazolo[4,5-*b*]pyridin3(2*H*)-yl)methyl]*O*,*O*-dimethyl phosphorothioate), and cypermethrin (cyano(3-phenoxyphenyl)methyl 3-(2,2-dichloroethenyl)-2,2-dimethylcyclopropanecarboxylate). The last three compounds in particular are highly toxic to aquatic invertebrates in laboratory tests, but their fate and effects in field situations has not been studied in detail.

7.9 Shipping

The volume of shipping has increased considerably during the past 40 years. Of environmental concern are discharges and emissions from vessels, and accidents. The International Maritime Organisation (IMO) administers the International Convention for the Prevention of Pollution from Ships (MARPOL 73/78, IMO World Wide Web site <url>www.imo.org</url>). The convention forbids discharges of "oily water" containing >100 mg/l of oil in proximity of land, and of any oil in particularly sensitive areas such as the Mediterranean. It also controls the pollution by assorted "noxious substance" carried in bulk as well as in packaged forms and freight containers, discharges generated by cleaning of chemical tankers, sewage, and garbage. The chemicals of concern are listed in Annex II of the convention [167]. No problems have been encountered in the North Sea as of 1989. A recent addition to MARPOL controls emissions from ships engine exhausts and limits the sulfur content of the fuel oil to 4.5% (1.5% in the Baltic).

Shipping accidents may result in spills of oil, bilge water, and loss of shipped materials. The frequencies of accidents range from 1×10^{-3} to 2×10^{-2} per ship per year. Accidents are more likely to occur in coastal waters [168] and those involving oils are twice as frequent as those involving chemicals. However, of the latter, chemicals carried in bulk are of greatest concern. As mentioned previously, the International Maritime Organisation (IMO) has a database of such chemicals.

7.10 Incineration at Sea

In the past, incineration was the disposal method of choice for organochlorine compounds such as PCBs, 2,4,5-trichlorophenoxyacetic acid (245 T), ethylene dichloride (EDC) tar, and chlorinated pesticides. It was used in Europe between 1969 and 1994, with a total of some 350 burns, and in the Gulf of Mexico and the Pacific Ocean, for 13 research burns between 1975 and 1983 [169]. Between 1980 and 1985, an estimated 100,000 tonnes of waste were incinerated annually, mostly in the North Sea (IMO web site), and regulations related to incineration at sea were added to the London Dumping Convention. The issue has caused considerable controversy, with strong opinions for and against [170–173]. Because of the risks of spills or other accidents at sea, and according to the precautionary principle, it seems that the practice has been abandoned world-wide [174] and the last incineration vessel was decommissioned in February 1991 (IMO web site).

7.11 Dumping of Wastes at Sea

Dumping of wastes at sea is internationally regulated by the London Dumping Convention (LDC) and its annual Consultative Meetings, administered by the International Maritime Organization (IMO). The history of LDC is described in a document available on the IMO World Wide Web site (<url>www.imo.org</url>). Developed in 1972, the LDC prohibits the dumping of organohalogen, mercury, and cadmium compounds, persistent plastics, oil and oil-based products, high-level radioactive wastes, and chemical and biological weapons (the "black list"). "Special care" is required when dumping wastes containing arsenic, lead, copper, zinc, and organosilicon compounds, as well as cyanides, fluorides, pesticides, and by-products. "Consideration shall be given" to wastes containing beryllium, chromium, nickel, and vanadium compounds (the "grey list"). LDC also lists various criteria that play a role in issuing dumping permits, such as the amount, form of waste (liquid, solid, sludge), etc., and the characteristics of the dumpsite.

About 80–90% of material dumped at sea are dredged sediments from maintenance of harbors and shipping channels. Of these, about 10% are heavily contaminated and their disposal is controlled by national regulations. The other materials dumped at sea include industrial waste (waste "generated by manufacturing or processing operations") and sewage sludge. According to IMO, the amount of industrial waste was 11–17 million tonnes per year in the 1970s and 8 million tonnes per year in the 1980s; the amount of sewage sludge was 17 million tonnes per year in the 1970s and 12 million tonnes per year in the 1990s, the latter mainly from Ireland, Japan, Republic of Korea, and the United Kingdom. In the North Sea, dumping of sewage sludge was supposed to cease at the end of 1998 [29]. In the US, 12–16 million tonnes per year of sewage sludge is dumped at specified marine disposal sites [175]. At a site that received 8 million tonnes from 1988 to1992, linear alkylbenzenes, coprostanol, and epi-coprostanol indicate that the sludge reached sediment at depths from 2400 m to 2900 m.

A 1990 amendment requires the consideration of the availability of scientific evidence for impact assessment before issuing a dumping permit. The Consultative Meeting also called for an end to dumping industrial waste by the end of 1995. In 1993, the Consultative Meeting also banned the dumping of low-level radioactive waste and changed the name of the Convention from LDC to London Convention (LC). The "precautionary approach", mentioned earlier, was adopted in 1996 in the "1996 Protocol". The protocol also prohibits the dumping of any waste except dredged material, sewage sludge, fish waste and material from industrial fish processing, vessels, and platforms, inert inorganic geological material, organic material of natural origin, and bulky items constructed of harmless materials, generated in locations which have no other means of disposal.

Only 76 countries have ratified the LC. Consequently, it is possible that uncontrolled dumping may be continuing in many other locations.

7.12 Other Industrial Effluents

Many industries are located on the sea coast, usually because of easy access to shipping, abundance of cooling water, sea-water as a raw material for the production of magnesium, effluent discharge, and dilution, etc. The main hazards of shipping are spills of oil or shipped materials and discharges of bilge and ballast water. Cooling water may be chlorinated to control fouling [176], may contain some corrosion inhibitors, or may be discharged at higher than ambient temperature. The latter may or may not be of advantage. Corrosion inhibitors may be organic compounds containing polar groups such as carboxyls or amino groups and non-polar carbon chains, which may be fluorinated. Inorganic corrosion inhibitors include phosphates, silicates, chromates, ferrous sulfate, or zinc salt and are frequently applied as mixtures of these compounds [177].

Chlorination of sea-water produces di- and trichloroacetic acid, as well as di- and tribromoacetic acid. Their concentrations are in the μg/l range and may not be environmentally significant. However, the halogenated acetic acids may lead to artifacts in measurements of organohalogen compounds in sea-water, and are of concern in drinking water produced on offshore platforms [178].

Effluents may be affecting the environment by the amounts of suspended solids such as effluents from phosphate fertilizer or titanium dioxide production or may contain one or, more usually, many chemicals with varying environmental impact. For example, 30 years ago, effluent from phosphorus production contained yellow phosphorus and caused spectacular kills of schools of herring that swam through the contaminated area, turned red by hemorrhaging, and died miles away from the exposure area [179]. Such surprises are much less likely now, but unanticipated chemicals or effects are likely to be encountered.

For complex effluents, particularly those from the chemical industry, the usual parameters such as BOD, COD, TOC, suspended solids, etc. are not sufficient characteristics and additional, more specific parameters are needed, such as the concentration of organically bound halogens, the concentrations of non-ionic, anionic and cationic surfactants, phenols, etc. The identification of chemicals may be guided by toxicity tests both in the laboratory and in the field [180].

Low-level chlorination of 0.5–1.5 mg/l, resulting in a chlorine residual of 0.1–0.2 mg/l, is used to reduce the degree of biofouling. In sea-water, chlorine oxidizes bromide, present at about 65 mg/l, to bromine, which also contributes to the generation of halogenated by-products [176]. The by-products include hypobromous acid, hypobromite, chloramines, bromamines, trihalomethanes, haloacetonitriles, haloacetic acids, and small amounts of halophenols. However, some haloforms and bromophenols as well as other organobromine compounds are also produced naturally in coastal waters [181, 182]. Empirical equations for the disappearance of chlorine/bromine derived oxidants from brackish water have been published [183].

8 Concluding Remarks

The amount of information on the presence in, and effects of chemicals on, the marine environment has increased tremendously during the past 40 years. Problems associated with sampling and sample handling are now well understood. Chemical analytical methods are much more sensitive, with limits of detection decreasing by three orders of magnitude per decade [184], and hyphenated methods providing increased selectivity. Many techniques for the measurement of biological effects are now available. International quality control and assurance protocols are implemented in monitoring the levels of chemicals and are being introduced into the monitoring of biological effects. There is also a considerable level of understanding of the principles of the behavior and effects of chemicals in the environment. All new chemicals or new uses of existing chemicals are carefully scrutinized and the possibility of an intentional introduction of new types of persistent and lipophilic organohalogen compounds into the environment is practically nil. Environmental impacts of new projects are considered beforehand and, quite frequently, the "precautionary principle" is applied.

At the same time, there is an inherent uncertainty quantified by the "Horwitz principle" in the measurements of trace concentrations, the extrapolation of laboratory toxicity studies to environmental situations is not straightforward, and unanticipated chemicals may occur in unexpected places or may display as yet unrecognized biochemical activities. The marine ecosystem is extremely complex and all the consequences of anthropogenic actions may never be fully predictable, but since the degree of predictability will increase with better information and its integration into knowledge and understanding, there is a continuing need for faster and cheaper methods of monitoring, evaluation, and information extraction from, and integration of, the data. At the same time, there is an urgent need to control and limit the amount of waste reaching the environment.

Acknowledgement. I thank Ms Joanne Cleghorn for fast, effective and resourceful assistance in obtaining literature.

9 References

1. Gianguzza A, Pelizzetti E, Sammartano S (eds) (1997) Marine chemistry. An environmental analytical chemistry approach. Kluwer, Dordrecht
2. Wade TL, Cantillo AY (1996) Use of standards and reference materials in the measurement of chlorinated hydrocarbon residues. Chemistry workbook, Intergovernmental Oceanographic Commission technical series 45, UNESCO, Paris
3. Foyn L (1997) Marine pollution. In: Brune D, Chapman DV, Gwynn MD, Pacyna JM (eds) The global environment. Science, technology and management, p 515
4. Wright JLC, Boyd RK, deFreitas ASW, Falk M, Foxall RA, Jamieson WD, Laycock MV, McCulloch AW, McInnes AG, Odense P, Pathak VP, Quilliam MA, Ragan MA, Sim PG, Thibault P, Walter JA, Gilgan M, Richard DJA, Dewar D (1988) Can J Chem 67:481
5. Diets R, Heide-Jorgensen MP, Harkonen T (1989) Ambio 18:258
6. Hall AJ, Law RJ, Wells DE, Harwood J, Ross HM, Kennedy S, Allchin CR, Campbell LA, Pomeroy PP (1992) Sci Total Environ 115:145
7. Sindermann CJ (1997) Mar Pollut Bull 34:218
8. MacGarvin M (1994) Precaution, science and the sin of hubris. In: O'Riordan T, Cameron J (eds) Interpreting the precautionary principle. Earthscan, London, p 69
9. Cole HA (ed) (1971) Proc Roy Soc Lond B 177:275
10. National Academy of Sciences (1971) Marine environmental quality – suggested research programs for understanding man's effect on the oceans. National Academy of Sciences, Washington, DC
11. GESAMP (1990) Review of potentially harmful substances – choosing priority organochlorines for marine hazard assessment. Rep Stud 42, Food and Agriculture Organization of the United Nations, Rome
12. GESAMP (1991) Global strategies for marine environmental protection. Rep Stud 45, International Maritime Organization, London
13. Jellinek, Schwartz, Connolly & Freshman, Inc and Coerr Environmental (1992) Environmental trends and issues at the research horizon – topical report. Gas Research Institute, Chicago
14. Goldberg ED (1995) Mar Pollut Bull 31:152
15. Krebs JR (1998) Our environmental future: the role of science. In: Lynch JM, Wiseman A (eds) Environmental biomonitoring: the technology ecotoxicology interface. Cambridge Univ Press, Cambridge, p 7
16. Høyer KG (1997) World in environmental transition. In: Brune D, Chapman DV, Gwynne MD, Pacyna JM (eds) The global environment. Science, technology and management, vol 1. Weinheim, Germany, p 7
17. Gatermann R, Huehnerfuss H, Rimkus G, Wolf M, Franke S (1995) Mar Pollut Bull 30:221
18. Key BD, Howell RD, Criddle CS (1997) Environ Sci Technol 31:2445
19. Goldberg ED (1976) The health of the oceans. The UNESCO Press, Paris
20. Blanck H, Dave G, Gustafsson K (1978) An annotated literature survey of methods for determination of effects and fate of pollutants in aquatic environments. National Swedish Environment Protection Board, Solna
21. GESAMP (1990) The state of the marine environment. UNEP Regional Seas Rep Stud 115, UNEP
22. HELCOM (1996) Third periodic assessment of the state of the marine environment of the Baltic Sea, 1989–1993. Background Document, Balt Sea Environ Proc 64B. Helsinki Commission, Helsinki

23. HELCOM (1997) Baltic marine environment bibliography 1993–1995, Balt Sea Environ Proc 66. Helsinki Commission, Helsinki
24. Marine Laboratory Aberdeen (1996) Environmental monitoring of the seas around Scotland 1970–1993. The Scottish Office, Agriculture, Environment and Fisheries Department, Aberdeen
25. Newman PJ, Agg AR (1988) Environmental protection of the North Sea. Heinemann Professional Publishing, Oxford
26. Wallace GT, Braash EF (eds) (1996) Proceedings of the Gulf of Maine ecosystem dynamics – a scientific symposium and workshop. RARGOM Rep 97–1, Hanover, NH
27. Eaton PB, Gray AG, Johnson PW, Hundert E (1994) State of the environment in the Atlantic Region. Environment Canada, Halifax, NS
28. White L, Johns F (1997) Marine environmental assessment of the estuary and Gulf of St. Lawrence. Fisheries and Oceans Canada, Dartmouth NS
29. North Sea Task Force (1993) North Sea quality status report 1993. Olsen & Olsen Fredensborg, Denmark
30. Siung-Chang A (1997) Environ Geochem Health 19:45
31. Farrington JW, Tripp BW, (eds) (1995). International mussel watch project, initial implementation phase. Final Rep NOAA Tech Memorandum NOS ORCA 95.US Department of Commerce National Oceanic and Atmospheric Administration, Silver Spring, Maryland
32. ICES (1997) Report of the ICES advisory committee on the marine environment 1997, Cooperative Research Report No 222. International Council for the Exploration of the Sea, Copenhagen
33. ICES (1989) Statistical analysis of the ICES cooperative monitoring programme. Data on contaminants in fish muscle tissue (1978–1985) for determination of temporal trend, Cooperative Research Rep No 162. International Council for the Exploration of the Sea, Copenhagen
34. ICES (1995) Report on the results of the ICES/IOC/OSPRCOM intercomparison programme on the analysis of chlorobiphenyls in marine media-step 2; and The intercomparison programme on the analysis of PAHs in marine media-stage 1, Cooperative Research Report No 207. International Council for the Exploration of the Sea, Copenhagen
35. Sylvester BA, Garton LS, Autenreich RL (1994) Water Environ Res 66:496
36. NOAA (1991) Second summary of data on chemical contaminants in sediments from the National Status and Trends Program, NOAA Technical Memorandum. NOS OMA 59, Rockville, MD
37. Daskalakis KD, O'Connor TP (1994) Inventory of chemical concentrations in coastal and estuarine sediments, NOAA Technical Memorandum NOS ORCA 76. National Oceanic and Atmospheric Administration, Seattle WA
38. Long ER, Field LJ, MacDonald DD (1998) Environ Toxicol Chem 17:714
39. Rowlatt SM, Davies IM (eds) (1995) Results of the 1990/1991 baseline study of contaminants in North Sea sediments, Cooperative Research Rep 208. International Council for the Exploration of the Sea, Copenhagen
40. Perttillae M, Bruegmann L (1992) Review of contaminants in Baltic sediments, Cooperative Research Rep 180. International Council for the Exploration of the Sea, Copenhagen
41. Wilkinson WB, Leeks GJL, Morris A, Walling DE (1997) Sci Total Environ 194/195:5
42. Fowler SW (1990) Mar Environ Res 29:1
43. Traunspurger W, Drews C (1996) Hydrobiologia 328:215
44. White HH (ed) (1984) Concepts in marine pollution measurements. University of Maryland, College Park
45. Reish DJ, Oshida PS, Mearns AJ, Ginn TC, Godwin-Saad EM, Buchman M (1997) Water Environ Res 69:877
46. Roig LC, Bergs MA (1997) Water Environ Res 69:727
47. Sutter GW II, Rosen AE (1988) Environ Sci Technol 22:548

48. Syversen T (1992) GESAMP composite list database, version 6. Department of Pharmacology and Toxicology, University of Trondheim, Trondheim
49. O'Connor TP, Daskalakis KD, Hyland JL, Paul JF, Summers JK (1998) Environ Toxicol Chem 17:468
50. Wolfe DA, Long ER, Thursby GB (1996) Estuaries 19:901
51. Chapman PM (1996) Ecotoxicology 5:327
52. Carr RS, Chapman DC, Howard CL, Biedenbach JM (1996) Ecotoxicology 5:341
53. Quevauviller P, Cofino W, Vijverberg F, Wells DE, Griepink B (1992) Quality assurance in marine monitoring, BCR Information Chemical Analysis. Commission of the European Communities, Luxembourg
54. Thompson M, Lowthian PJ (1997) J AOAC Int 80:676
55. Grasselli JG (1992) Anal Chem 64:677A
56. QUASIMEME Project office (1998) QUASIMEME laboratory performance studies June 1996-May 1997, QUASIMEME Bulletin No. 5. QUASIMEME Project Office, Aberdeen
57. Allan A (1998) Quality assurance of sampling and sample handling (QUASH) launched at RIKZ, Groningen. In: QUASIMEME Bulletin 5. QUASIMEME Project Office, Aberdeen, p 11
58. Breen JJ, Robinson PE (eds) (1985) Environmental applications of chemometrics, ACS Symposium Series 292. American Chemical Society, Washington, DC
59. Sharaf MA, Illman DL, Kowalski BR (1986) Chemometrics. Wiley, New York
60. Massart DL, Vandeginste BGM, Deming SN, Michotte Y, Kaufman L (1988) Chemometrics: a textbook. Elsevier, Amsterdam
61. Malinowski ER (1991) Factor analysis in chemistry. Wiley, New York
62. Grant A (1990) Mar Pollut Bull 21:297
63. Adami G, Aleffi F, Barbieri P, Favretto A, Predonzani S, Reisenhofer E (1997) Water Air Soil Pollut 99:615
64. Corbella R, Garcia-Montelongo F (1997) Quim Analit 16(Suppl 2):275
65. Sicre MA, Paillasseur JL, Marty JC, Saliot A (1988) Org Geochem 12:281
66. Wenning RJ, Harris MA, Ungs MJ, Paustenbach DJ, Bedbury H (1992) Arch Environ Contam Toxicol 22:397
67. Wenning RJ, Paustenbach DJ, Harris MA, Bedbury H (1993) Arch Environ Contam Toxicol 24:271
68. Younker MB, Cretney WJ (1996) Dioxins and furans in crab hepatopancreas: use of principal component analysis to classify congener patterns and determine linkages to contamination sources. In: Servos MR, Munkittrick KR, Carey JH, van der Kraak GJ (eds) Environmental fate and effects of pulp and paper mill effluents. St Lucie Press, Delray Beach, Florida, p 315
69. Kucuksezgin F (1996) Toxicol Environ Chem 55:135
70. Salau JSI, Tauler R, Bayona JM, Tolosa I (1997) Environ Sci Technol 31X:3482
71. Lithner G, Broman D, Naef C, Borg H, Johansson AM, Kaerrhage P, Larsson MB (1996) Metals in settling particles and surficial sediments of the Swedish Baltic coast 1988–1989. In: Munawar M, Dave G (eds) Development and progress in sediment quality assessment: rationale, challenges, techniques & strategies. SPB, Amsterdam, p 27
72. Gemperline PJ, Miller KH, West TL, Weinstein JE, Hamilton JC, Bray JT (1992) Anal Chem 64:523A
73. Devillers J, Domine D, Garrigues P, Budzinski H, Karcher W (1996) Polycyclic Arom Compds 11:219
74. Beliaeff B, O'Connor TP, Daskalakis DK, Smith PJ (1997) Environ Sci Technol 31X:1411
75. Haahti H, Perttilae M (1988) Mar Pollut Bull 19:29
76. Loganathan BG, Kannan K (1991) Mar Pollut Bull. 22:582
77. Nicholson MD, Fryer RJ, Ross CA (1997) Mar Pollut Bull 34:821
78. Stagg RM (1998) Quality assurance and biological effects measurement. In: QUASIMEME Bulletin 5. QUASIMEME Project Office, FRS Marine Laboratory, Aberdeen, p 17
79. Sinderman CJ (1995) Ocean pollution – effects on living resources and humans. CRC Press, Boca Raton

80. McCain BB, Brown DW, Krahn MM, Myers MM, Clark RC Jr, Chan S-L, Malins DC (1988) Aquat Toxicol 11:143
81. Stehr CM, Myers MS, Burrows DG, Krahn MM, Meador JP, McCain BB, Varanasi U (1997) Ecotoxicology 6:35
82. Johnson LL, Stehr CM, Olson OP, Myers MS, Pierce SM, Wigren CA, McCain BB, Varanasi U (1993) Environ Sci Tech 27:2759
83. Pedersen B, Cofino W, Davies I (1997) Mar Pollut Bull 35:42
84. Senten JR, Charlier RH (1991) Int J Environ Studies 37:79
85. Cantillo AY, O'Connor TP (1992) Chem Ecol 7:31
86. Nakamura Y, Tokunaga T (1996) Wat Sci Tech 34:133
87. Bryan GW, Langston WJ (1992) Environ Pollut 76:89
88. Neff JM (1997) Environ Toxicol Chem 16:917
89. Farago ME (1997) Arsenic in the marine environment. In: Gianguzza A, Pelizzetti E, Sammartano S (eds) 1997 Marine chemistry. An environmental analytical chemistry approach. Kluwer, Dordrecht, pp 275
90. Lemly DA (1998) Ecotoxicol Environ Safety B 39:1
91. Wendt PH, Van Dolah RF, Bobo MY, Mathews TD, Levisen MV (1996) Arch Environ Contam Toxicol 31:24
92. GESAMP (1990) Review of potentially harmful substances. Nutrients Rep Stud GESAMP 34
93. Viviani R (1992) Sci Total Environ, Supplement 631
94. Aminot A, Kirkwood D, Carlberg S (1998) Mar Pollut Bull 35:28
95. Konings H (1996) Spill Sci Technol Bull 3:47
96. Lindgaard-Jorgensen P, Bender K (1994) Wat Sci Tech 29:165
97. GESAMP (1993) Impact of oil and related chemicals and wastes on the marine environment. Rep Stud GESAMP 50
98. Krahn MM, Stein JE (1998) Anal Chem 70:186A
99. Tahir NM, Abdullah AR, Shanmugam S (1997) Environ Geochem Health 19:67
100. Hudgins CM Jr (1991) Chemical usage in North Sea oil and gas production and exploration operations. A report prepared for the Norwegian Oil Industry Association. Petrotech Consultants, Houston
101. Hudgins CM Jr (1992) J Pet Technol 44:604
102. Vik EA, Berg JD Bakke S, Lundh T (1991) Wat Sci Tech 24:135
103. Vik EA, Bakke S, Johnson DR, Verburgh J (1996) The *n*-octanol/water partition coefficient. A critical parameter in environmental risk assessment of offshore E&P chemicals. In: Reed M, Johnsen S (eds) Produced water 2: environmental issues and mitigation technologies. Plenum Press, New York, p 135
104. Bjornestad E, Petersen GI, Robson M, Reiersen LO, Henriquez L, Massie L, Blackman R (1993) Sci Total Environ (Suppl):713
105. Webster L, Mackie PR, Hird SJ, Munro PD, Brown NA, Moffat CF (1997) Analyst 122:1485
106. Zitko V (1993) Cent Eur J Publ Health 1:125
107. Zitko V, Lindsay G (1995) Toxicol Modelling 1:35
108. Naes K, Oug E (1997) Environ Sci Technol 31:1253
109. Lipiatou E, Tolosa I, Simo R, Bouloubassi I, Dachs J, Marti S, Sicre MA, Bayona JM, Grimalt JO, Saliot A, Albaiges J (1997) Deep-Sea Res II 44:881
110. Gustafsson O, Gschwend PM (1997) Soot as a strong partition medium for polycyclic aromatic hydrocarbons in aquatic systems. In: Eganhouse RP (ed) Molecular markers in environmental geochemistry, ACS Symposium Series 671. ACS Washington DC, p 365
111. Naes K, Oug E, Knutzen J (1998) Mar Environ Res 45:193
112. Webster L, Angus L, Topping G, Dalgarno EJ, Moffat O (1997) Analyst 122:1491
113. Meador JP, Stein JE, Reichert WL, Varanasi U (1995) Rev Environ Contam Toxicol 143:79
114. Tuvikene A (1995) Ann Zool Fennici 32:295
115. Horness BH, Lomax DP, Johnson LL, Myers MS, Pierce SM, Collier TK (1998) Environ Toxicol Chem 17:872
116. Sved DW, Roberts MH Jr, van Veld PA (1997) Wat Res 31:294

117. Law RJ, Klungsoyr J, Freriks IL (1997) Mar Pollut Bull 35:64
118. Xie WH, Shiu WY, Mackay D (1997) Mar Environ Res 44:429
119. Delbeke K, Zhu L, Muqtadir M, Khaled M (1995) Intern J Anal Chem 58:131
120. Young RA, Swift DJP, Clark TL, Harvey GR, Betzer PR (1985) Science 229:431
121. Loganathan BG, Kannan K (1994) Ambio 23:187
122. Tanabe S, Iwata H, Tatsukawa R (1994) Sci Total Environ 154:163
123. Westgate AJ, Muir DCG, Gaskin DE, Kinglsey MCS (1997) Environ Pollut 95:105
124. Langston WJ (1995) Pestic Outlook Dec:18
125. Scarlett A, Donkin ME, Fileman TW, Donkin P (1997) Mar Pollut Bull 34:645
126. Davies IM, Bailey SK, Harding MJC (1998) ICES J Mar Sci 55:34
127. Buser HR, Mueller MD, Theobald N (1998) Environ Sci Technol 32:188
128. Bester K, Huehnerfuss H (1996) Chemosphere 32:1919
129. Law RJ, de Boer J (1995) Quality assurance of analysis of organic compounds in marine matrices: application to analysis of chlorobiphenyls and polycyclic aromatic hydrocarbons. In: Quevauviller P (ed) Quality assurance in environmental monitoring. Wiley-VCH, Weinheim, p 129
130. Kodavanti PRS, Tilson HA (1997 Neurotoxicol 18:425
131. Anonymous (1979) Polychlorinated biphenyls. National Academy of Sciences, Washington, DC
132. Tolosa I, Readman JW, Fowler SW, Villeneuve JP, Dachs J, Bayona JM, Albaiges J (1997) Deep-Sea Res II 44:907
133. Naef C, Axelman J, Broman D (1996) Organic contaminants in sediments of the Baltic Sea: distribution, behaviour and fate. In: Munawar M, Dave G (eds) Development and progress in sediment quality assessment: rationale, challenges, techniques & strategies. Academic Publishing, Amsterdam, p 15
134. De Boer J, Wells DE (1997) Mar Pollut Bull 35:52
135. Hope B, Scatolini S, Titus E, Cotter J (1997) Mar Pollut Bull 34:548
136. Falandysz J, Strandberg B, Strandberg L, Bergqist PA, Rappe C (1997) Sci Total Environ 204:97
137. Jaernberg U, Asplund L, de Wit C, Egebaeck AL, Wideqvist U, Jakobson E (1997) Arch Environ Contam Toxicol 32:232
138. Kurz J, Ballschmiter K (1995) Fresenius J Anal Chem 351:98
139. Engwall M, Broman D, Naef C, Zebuehr Y, Brunstroem B (1997) Mar Pollut Bull 34:1032
140. Haglund PS, Zook DR, Buser HR, Hu J (1997) Environ Sci Technol 31:3281
141. De Boer J, Wester PG, Evers EHG, Brinkman UATh (1996) Environ Pollut 93:39
142. Mu H, Wesen C, Sundin P (1997) Trends Anal Chem 16:266
143. Boyle JL, Lindsay RC, Stuiber DA (1992) J Food Sci 57:918
144. Fukushima M, Kawai S, Yamaguchi Y (1992) Wat Sci Technol 25:271
145. Hale RC, Smith CL (1996) Intern J Environ Anal Chem 64:21
146. Redford D, Trulli WR, Trulli HK (1992) Chem Ecol 7:75
147. Zitko V (1993) Mar Pollut Bull 26:584
148. Redford D, Billy M, Tom A (1992) Chem Ecol 6:189
149. Eganhouse RP, Olaguer DP, Gould BR, Phinney CS (1988) Mar Environ Res 25:1
150. Granmo A, Kollberg S, Berggren M, Ekelund R, Magnusson K, Renberg L, Wahlberg C (1991) Bioaccumulation of nonylphenol in caged mussels in an industrial coastal area on the Swedish west coast. In: Angeletti G, Bjorseth A (eds) Organic micropollutants in the aquatic environment, Proc 6th European Symposium, Lisbon, Portugal May 22–24, 1990, p 71
151. Stalmans M, Matthijs E, de Oude NT (1991) Wat Sci Tech 24:115
152. Gonzales-Mazo E, Honing M, Barcelo D, Gomez-Parra A (1997) Environ Sci Technol 31:504
153. Preston MR (1997) Sewage and nutrients in the marine environment: stimulants for good or vectors for harm? In: Gianguzza A, Pelizzetti E, Sammartano S (eds) Marine chemistry. An environmental analytical chemistry approach. Kluwer, Dordrecht, p 259
154. Gay J, Webster R, Roberts D, Trett M (1991) J Inst Water Environ Manag 5:573

155. Theodorou AJ (1997) Fresenius Environ Bull 6:397
156. Valls M, Bayona JM, Albaiges J (1989) Nature 337:722
157. Colodey AG, Wells PG (1992) J Aquat Ecosyst Health 1:201
158. Enell M (1996) Load from the Swedish pulp and paper industry (nutrients, metals and AOX): quantities and shares of the total load on the Baltic Sea. In: Servos MR, Munkittrick KR, Carey JH, van der Kraak GJ (eds) Environmental fate and effects of pulp and paper mill effluents. St Lucie Press, Delray Beach, Florida, p 229
159. Servos MR, Munkittrick KR, Carey JH, van der Kraak GJ (eds) (1996) Environmental fate and effects of pulp and paper mill effluent. St Lucie Press, Delray Beach, Florida
160. Wulff F, Rahm L, Jonsson P, Brydsten L, Ahl T, Granmo A (1993) Ambio 22:27
161. Mayer I, McLean E (1995) Wat Si Tech 31:85
162. Wu RSS (1995) Mar Pollut Bull 31:159
163. Zitko V (1994) Chemicals in aquaculture (an overview). In: Ervik A, Hansen PK, Wennevik V (eds) Proceedings of the Canada-Norway workshop on environmental impacts of aquaculture, Fisken og Havet 13. Institute of Marine Research, Bergen, p 97
164. Findlay RH, Watling L, Mayer LM (1995) Estuaries 18:145
165. Panchang V, Cheng G, Newell C (1997) Estuaries 20:14
166. Enell M (1995) Wat Sci Tech 31:61
167. Hurford N, Law RJ, Payne AP, Fileman TW (1989 Oil & Chem Pollut 5:391
168. Romer H, Brockhoff L, Haastrup P, Petersen HJS (1993) J Loss Prevent Process Indust 6:219
169. Ditz DW (1988) J Haz Mat 17:149
170. Nassos GP (1987) Mar Pollut Bull 18:211
171. Spaans L (1988) Mar Pollut Bull 19:256
172. Bond DH (1984) Environ Sci Technol 18:148A
173. Bond DH (1985) Environ Sci Technol 19:486
174. Kasoulides GC (1988) Mar Pollut Bull 19:648
175. Takada H, Farrington JW, Bothner MH, Johnson CG, Tripp BW (1994) Environ Sci Tech 28:1062
176. Jenner HA, Taylor CJL, van Donk M, Khalanski M (1997) Mar Environ Res 43:279
177. White R (1989) Corrosion and Coatings South Africa 16:3
178. Kristiansen NK, Aune KT, Froshaug M, Becher G, Lundanes E (1996) Wat Res 30:2155
179. Jangaard PM (ed) (1972) Effects of elemental phosphorus on marine life. Collected papers resulting from the 1969 pollution crisis, Placentia Bay, Newfoundland, Circular 2. Fisheries Research Board of Canada, Halifax, NS
180. Nyholm N (1992) Wat Sci Tech 25:449
181. Butler A, Walker JV (1993) Chem Rev 93:1937
182. Gribble GW (1994) Environ Sci Technol 28:310A
183. Yamamoto K, Fukushima M (1992) Wat Res 26:1105
184. Harris WE (1992) Anal Chem 64:665A

CHAPTER 5

Dissolved Organic Carbon from Phytoplankton

Sverre M. Myklestad

Department of Biotechnology, Norwegian University of Science and Technology, NTNU, 7491 Trondheim, Norway
E-mail: Sverre.Myklestad@chembio.ntnu.no

Dissolved organic carbon in the sea makes up some 700 gigatons according to conservative estimates. Even more important is the labile part, which constitutes more than 50% of newly produced DOC. The source of DOC production in the sea is photosynthetic algae. The major constituent in rapidly growing phytoplankton cells is protein, with a content up to about 50% of organic dry matter; the content of carbohydrate and lipid being somewhat lower. When growth is slow and in stationary phases, the cellular composition will change very markedly and often carbohydrate will be the most prominent constituent.

Extracellular release is now well established as a part of the primary production. Rapidly growing phytoplankton releases 2–10% (PER) in most cases, increasing to 10–60% in the stationary phase of growth mainly because of a lower photosynthetic rate. There is increasing evidence that the absolute rate of release is highest in the exponential phase of growth. Nutrient limitation has been shown to increase the relative rate of release and nutrient ratios, i.e. N/P, also seems to be of importance. Extracellular release is relatively unaffected by irradiance but PER is correlated to the relative inhibition of photosynthesis at high irradiances.

Simple diffusion has been suggested to be the important mechanism for transporting small molecules through the cell membrane, and transport of large molecules is most likely mediated by transport vesicles loaded with biopolymers budding off from donor membranes and fusion to acceptor membranes before delivering the load to the external medium.

Leaky cells represent a completely different cause of release of organic compounds. A number of physical stress factors in the environment, such us severe nutrient limitation and far from optimal conditions of temperature, salinity, pH, and light may lead to unhealthy cells and finally cell death i.e. cell lysis. Biological factors such as the attack of viruses, bacteria and possibly heterotrophic flagellates may lead to the same fatal result. Rapidly growing cells and stationary phase cells, respectively, may contain approximately 15–30% and 20–50% of cell carbon as soluble compounds composed of small molecules, soluble proteins and carbohydrate storage compounds. During lysis, therefore, the cell will loose significant amounts of cell material to the external medium.

Another kind of leakage is inefficient feeding by zooplankton leading to loss of dissolved and suspended material.

Bacteria have a great tendency to attach to non-growing or slow-growing phytoplankton cells. Degrading enzymes on bacterial surfaces catalyze degradation of phytoplankton cellular biopolymers. The bacteria eventually take up some of the products and the fraction of soluble compounds not taken up is released.

Average values from a number of field investigations recorded show an approximate range in PER of 5–30, which is significantly higher than the release from exponentially growing cells in culture. The importance of the taxonomic composition of the phytoplankton has now been shown not only in culture but also in natural populations.

While protein is the major cellular component in exponentially growing cells, carbohydrate is the most prominent substance among the extracellular products, with protein and amino acids probably in second place. Other important substances to mention are organic acids, sugar alcohols, lipids and fatty acids, vitamins and growth inhibiting compounds (toxins). The list could have been made much longer.

The Handbook of Environmental Chemistry Vol. 5 Part D
Marine Chemistry
(ed. by P. Wangersky)

Special events described involving extracellular production as an important element, occurring in the Adriatic Sea, the southern part of the North Sea and Scandinavian waters illustrate cases where these products may strongly influence the entire local environment.

Keywords: DOC, Excretion, Chemical composition, Lysis, Adriatic Sea, Physiological mechanisms, Phytoplankton.

1 Introduction

Dissolved organic carbon is a huge carbon pool in the sea making up some 700 gigatons [1] according to conservative estimates. The concentration is not constant throughout the oceans; rather changing throughout the year, according to depth and latitude. The DOC in deep water is old (3000–6000 years) and refractory [2]. This points to the fact that a large part of the DOC in the ocean is very fragile. However, in the newer literature, many reports have dealt with the labile part, which may vary from a few to perhaps 35% of the total DOC [3–5].

The origin of DOC is photosynthetic organic material from phytoplankton as this algal group is by far the most important primary producer. In principle, all cellular and extracellular compounds could appear as soluble compounds although only after degradation of most of the cellular biopolymers. The material is released to the water as extracellular products of photosynthesis, as leakage from cells caused by different processes, and also by solubilization by degradation of cellular, polymeric material.

DOC is defined more operationally than absolutely, and is often referred to as the fraction passing through a 0.5-µm filter [6] or a GF/F glass-fiber filter [7, 8]. More interesting than the absolute amounts is the chemical composition, which generally is very complex, and less than 25% can be accounted for as known substances. The most important compounds released from phytoplankton are carbohydrates (mono- and polysaccharides), proteins, amino acids, organic acids, and vitamins. One of the most serious constraints to more data and new information is still the availability of sensitive and precise methods. This review has drawn upon a large number of laboratory and field investigations and some effort is made to highlight some papers of special significance in the development of thinking and research in the field. As a part of this, different mechanisms for release of organics from phytoplankton will be treated. Although not a main question here, interaction between DOC and the environment will be considered.

The paper as a whole focuses on DOC from phytoplankton in terms of quantitative amounts and characteristics of the material, the conditions and mechanisms of release to the water, and touches upon its interaction with the environment.

2 Photosynthetic Production and Chemical Composition of Phytoplankton

In order to get a better understanding of the production of dissolved organic compounds, it may be useful to look at the general synthesis of organic material from our algal primary producers.

The first product of the CO_2 reduction process in photosynthesis is 3-phosphoglycerate, which then is reduced and converted to glyceraldehyde-3-phosphate. This is a real 3-carbon carbohydrate which is further metabolized in the glycolytic pathway and the TCA-cycle, or may follow the anabolic pathway to glucose-1-phosphate. Precursors from the glycolysis and the TCA-cycle are the starting material for the cell to make amino acids and proteins, mono-and polysaccharides, fatty acids and lipids, nucleotides and the nucleic acids DNA and RNA, just to mention a few of the quantitatively most important of at least more than a thousand different organic molecules in one single cell.

Phytoplankton cells have a fantastic cellular machinery to make all the intricate molecules from simple inorganic nutrients. Growth is not always optimal

Table 1. Content of protein, carbohydrate, lipid and ash in phytoplankton cultures as percentages of ash-free dry weight

Species	Protein	Carbohydrate	Lipid	Ash	Source
Chlorophyceae					
Tetracelmis maculata	68	19.7	3	23.8	Parsons et al., 1961
Dunaliella salina	62	34.1	3.3	7.6	Parsons et al., 1961
Chlorella protothecoides[a]	26	11	13	–	Brown and Jeffrey, 1992
Prymnesiophycea					
Monochrysis lutheri	52	33.6	12.3	6.4	Parsons et al., 1961
Syrachosphaera carterae	88	28	7.2	36.5	Parsons et al., 1961
Pavlova lutheri (exp.)	34	18.5	29	55	Fernandez-Reiriz et al., 1989
Pavlova lutheri (early stat)	43	15.6	29	35.5	Fernandez-Reiriz et al., 1989
Pavlova lutheri (late stat.)	16	53	38	14	Fernandez-Reiriz et al., 1989
Bacillariophycea					
Chaetoceros sp.	49	9.2	9.6	28	Parsons et al., 1961
Skeletonema costatum	61	34.1	7.7	39	Parsons et al., 1961
Coscinodiscus sp.	40	9.5	4.2	57	Parsons et al., 1961
Coscinodiscus eccentricus[b] (exp.)	60	12.9	9	–	Pugh,1975
Coscinodiscus eccentricus (stat.)	41	9.2	19.6[c]		Pugh,1975
Skeletonema costatum (exp.)	41	48			Myklestad, 1974
Skeletonema costatum (early stat.)	31	53			Myklestad, 1974
Skeletonema costatum (late stat.)	11	90			Myklestad, 1974
Corethron hystrix (exp.)	49	12.8			Myklestad, 1974
Corethron hystfix (stat.)	18.5	63	–		Myklestad, 1974
Dinophyceao					
Amphidinium carteri	33	41	20.9	14.1	Parsons et al., 1961
Exuviella sp.	34	40	16.3	8.3	Parsons et al., 1961

[a] % of dry weight.
[b] 30 PSU.
[c] Calculated by difference.

and is sensitive to changes in nutrient concentration, light, temperature, salinity, pH and to physical properties of the water masses such as turbulence. Growth conditions will also influence internal biochemistry and the release of organic material from phytoplankton. Healthy cells will release a small part of their photosynthetic production extracellularly as a normal process. If the cells becomes leaky for some reason, they may lose some or all of their soluble compounds and degradation processes effected by enzymatic and chemical reactions may further solubilize cell constituents. Cellular degradation processes are going on all the time, giving many constituents their characteristic turnover times. These products will be part of the cell's soluble material as well.

Table 1 shows the major constituents of cultured microalgal cells from different classes as a percentage of ash-free dry weight together with ash content [9–13]. The chemical composition of algae is often dissimilar for different classes and species. It varies markedly as a function of growth conditions and is particularly sensitive to nutrient limitation [9]. Protein content varies from about 35–50%, carbohydrate from 10–40% and lipid from 5–30% in the exponential phase of growth, while in the stationary phase protein may decrease down to 15%, carbohydrate increase up to 70–90% in some cases and the content of lipid may increase to 35–40%. The ash content may vary from 6–60% with the highest values in diatoms.

Fernandez-Reiriz [10] gave a range of 0.3–2% for RNA in seven different species. The DNA content is generally lower and has been reported in the range 0.2–4.7% of the protein level [14–15]. DNA-N was found to vary from 0.4–4.2% and RNA from 1.1–9.8% of total cellular-N in four species of phytoplankton [16]. When focusing on release to the environment, soluble cellular organic constituents are especially interesting because of the potentially fast release of this fraction of the cell carbon. The principal soluble compounds are a part of the cellular carbohydrates, free amino acids, soluble proteins, free fatty acids and often small amounts of sugar alcohols, organic acids, nucleotides, vitamins and many other metabolites.

3
Release of Dissolved Organic Carbon

Phytoplankton cells are unique organisms that make all of the intricate components necessary to build cells from simple inorganic nutrients. Populations are growing at a rate of about one half to three divisions per day; all at the expense of solar energy. Ideally one should think that all reduced carbon products from the photosynthesis should remain in the cells. This absolute efficiency is not the case in nature and algal cells are also producing extracellularly in addition to the main cellular photosynthetic production. The appearance of photosynthetic products in the medium of a healthy culture is defined as extracellular production; although healthy cells may not necessarily grow at their maximum specific growth rate [17].

In the scientific literature, extracellular production and related processes have many names: excretion, exudation, release, and liberation. I will often use release and excretion in addition to extracellular production.

Three important developments during the last 20 years or so have particularly influenced the research activity in the field of release of dissolved organic carbon:

1. A critical discussion of the process of excretion and the methods involved in its investigation – in particular the ^{14}C-procedures [18–21].
2. Improvement of existing and development of new chemical methods for the quantitative determination of soluble components [136].
3. The importance of exudates from phytoplankton as a source of nutrients for microheterotrophs in the sea.

Sharp (1977) made a careful examination of earlier work and concluded that an accumulation of organic matter was taking place in the stationary phase of growth, but excretion during active growth was not appreciable. Potential methodological problems were part of his concern, e.g. cell rupture during manipulation, residual ^{14}C inorganic carbon and organic contamination of ^{14}C. His work was extremely important because he placed a question mark on excretion by marine phytoplankton: "Do healthy cells do it?" The immediate replies of Fogg [19] and Aaronson [20] were clarifying, but even more important was a renewed interest in the field from several workers.

Table 2 shows some values of release in exponential and stationary phase cultures of marine phytoplankton [22–23; 26–28; 30–31; 37; 40; 133]. It is important to define quantitative measures of extracellular production: Total photosynthetic production = cellular + extracellular production. Relative extracellular release as percentage (PER) = extracellular production × 100/total photosynthetic production.

Absolute extracellular release may be given as the quantity of C or a particular constituent per cell or unit of biomass per time unit (e.g. pg C. $cell^{-1} day^{-1}$).

Rapidly growing cultures (Table 2, exp. phase) show PER values from 2–10 in most cases, but may increase to 10–30 in some cases. Mague et al. [22] concluded that extracellular release is a normal function of healthy cells and showed that the relative composition of *Skeletonema costatum* cellular and the extracellular fraction of amino acids was quite different. Marlowe et al. [23] fractionated cellular and extracellular material and found a marked difference in the content of ionic compounds in the two fractions. Myklestad et al. [24] found the reserve polysaccharide β-1,3-glucan to be quite dominating in the soluble cellular fraction of *Chaetoceros affinis* while the extracellular polysaccharide produced was entirely different and contained no glucose at all in the molecule [25]. These investigations show clearly that extracellular production observed was not just a portion of the cell content and therefore not attributable to cell leakage or lysis. Myklestad et al. [26] also measured release rates of carbohydrate and amino acids from *C. affinis* and concluded that extracellular production is taking place in all growth phases in culture (Table 3). It should be noted that absolute release per cell was as high or higher in the exponential phase than in the stationary phase.

The level of release according to Table 2 is from 2–10% and approaching 30% in one case in the exponential phase of growth, and from about 10–60% in the stationary phase. The reason for this large difference in percentage re-

Table 2. Extracellular release of soluble organic carbon as percentage of total photosynthetic production (PER) for phytoplankton cultures

Author	Year	Organism	PER Growth phase Exp.	Stat.	Method
Hellebust	1965	22 species	3–6[a]	–	^{14}C tracer technique
Hellebust	1967	*T. fluviatilis*	5	20	^{14}C tracer technique
Berman & Holm-Hansen	1974	8 species	5	–	^{14}C tracer technique
Laws & Wong	1978	3 species	1.5	–	^{14}C tracer technique
Mague et al.	1980	*S. costatum*	8	–	^{14}C; chem. methods
Claus	1988	*S. costatum* *C. radians*	2–8	11–39	Chem. methods
Marlowe et al.	1989	*Hymenomonas carterea*	4–10	20–64	^{14}C; chem. methods
Myklestad et al.	1989	*C. affinis*	10	58	Chem. methods
Zlotnik & Dubinsky	1989	3 species	< 10	–	^{14}C tracer technique
Obernosterer & Herndl	1995	*C. affinis*	15–29	29–37	^{14}C; chem. methods

[a] Of 22 species from 5 algal classes, a few excreted 10–25%.

lease is the sharp decline in photosynthetic production in a culture after reaching the stationary phase. The absolute release may not be higher during the stationary phase in spite of many reports claiming this. Mague et al. [22] put forth the notion that not only PER should be reported, but also absolute rates of cellular and dissolved production. This will definitely give a more complete picture of the magnitude and nature of extracellular release.

Carluzzi and Bowes [29] showed a completely different approach to the very existence of extracellular production of healthy cells and gave a beautiful demonstration of production and utilization of vitamins in cultures with 2 or more phytoplankton species present in the same flask. Thiamin, vitamins B_{12}, and biotin were all released, but from different algal species.

Sometimes large differences in total excretion between species and genera have been shown [30]. Within one algal class, diatoms, the author found excretion of glycolic acid varying from 1–35% of total excretion. Zlotnik and Dubinski [31] observed, that in *Isochrysis galbana*, the absolute excretion was temperature independent, while 2 other species showed such a dependency. All 3 species revealed temperature-dependent photosynthesis. Dramatic differences were demonstrated between species of marine diatoms in production of extracellular polysaccharides in stationary phase cultures [9]. Fig. 1 shows that extracellular polysaccharides as percentage of total cellular carbohydrate + extracellular carbohydrate varies from 1% in *Skeletonema costatum* to 56% in *Chaetoceros curvisetus*. It may be significant that the genera Chaetoceros (Bacillariophyceae) have a relatively higher extracellular production than other genera also when compared with several other algal classes.

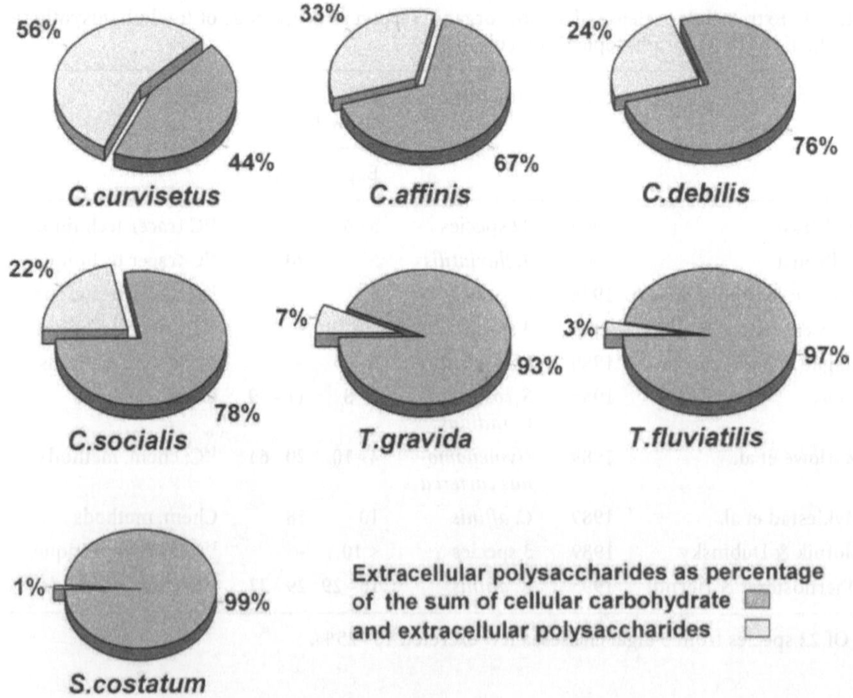

Fig. 1. Extracellular polysaccharide production in 12-day cultures for marine diatoms (*Chaetoceros, Thalassiosira* and *Skeletonema*) [9]

3.1 Nutrient Status

Cells growing rapidly have often been said to release only small amounts of extracellular organic material [18; 32; 33] although exceptions have been reported [34] where Hellebust refers to work with exponentially growing cells excreting from 1–50% of their photosynthetic production, depending on species and growth conditions. The excretion of glycolate by green algae during photosynthesis is well established in most classes of algae [30]. The excretion of this simple metabolite is much higher in rapidly growing cultures of *Chlamydomonas reinhardtii* than in older ones [35].

Nalewajko [38] has shown the same phenomenon very clearly with cultures of *Chlorella pyrenoidosa* (Fig. 2). There is a rapid release of glycolate during the exponential phase, but it stops abruptly when the culture reaches the stationary phase of growth. Zlotnik and Dubinsky [31] observed a tight coupling between rates of carbon fixation and of absolute DOC excretion over a wide range of irradiances using 2 fresh water green species and the marine *Isochrysis galbana*. Myklestad et al. [26] (Table 3) showed that the specific release of carbohydrates and amino acids for the marine diatom *Chaetoceros affinis* are significant in the exponential phase of growth and apparently higher than in the stationary

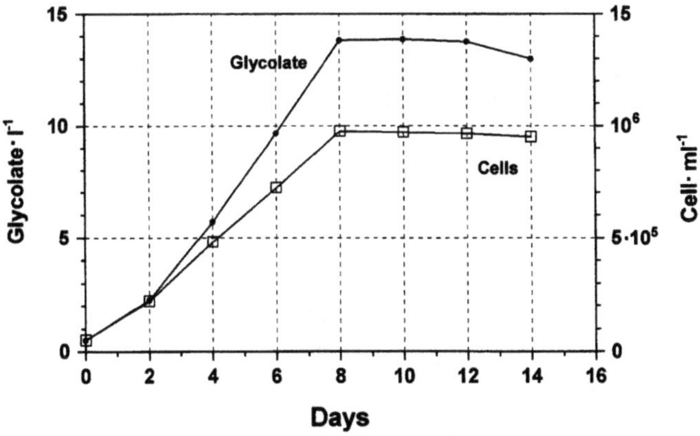

Fig. 2. Growth and glycolate excretion by axenic *Chlorella pyrenoidosa* [38]

phase. These examples, and the list is certainly not complete, may teach us that rapidly growing cultures with an ample supply of nutrients and otherwise optimal conditions may excrete dissolved organic matter to the medium at a relatively high absolute rate. Previously this has sometimes been "hidden" by the fact that many cultures under stress of different kinds showed much higher relative rates (PER).

There are numerous reports on the release of organic matter under limiting nutrient conditions [18; 33; 36–37]. A predominant factor why many observations have been made under low nutrient levels is that organic material, especially carbohydrates, have accumulated in the culture medium [9].

There has been an interest in looking at not only the effect of low and limiting nutrient concentrations, but also nutrient ratios. Since N and P are the major nutrients which may potentially limit phytoplankton primary production, the N/P ratio has been changed in culture work. One should also remember that this ratio really varies in many coastal- and near- shore waters. Myklestad and Haug [36] cultured the marine diatom *C. affinis* in media with different N/P ratios. Figure 3 shows the effect on the production of extracellular polysaccharide in 12-day cultures. Since a clear indication of higher production was seen in

Table 3. Rate of extracellular release of carbohydrate and free amino acids per cell in a *Chaetoceros affinis* culture for different growth phases [26]

Growth phase	Rate of extracellular release	
	Carbohydrate (pg cell^{-1} day^{-1})	Amino acids pmol cell^{-1} day^{-1}
Exponential	42	0.038
Transition	18	0.025
Stationary	20	0.026

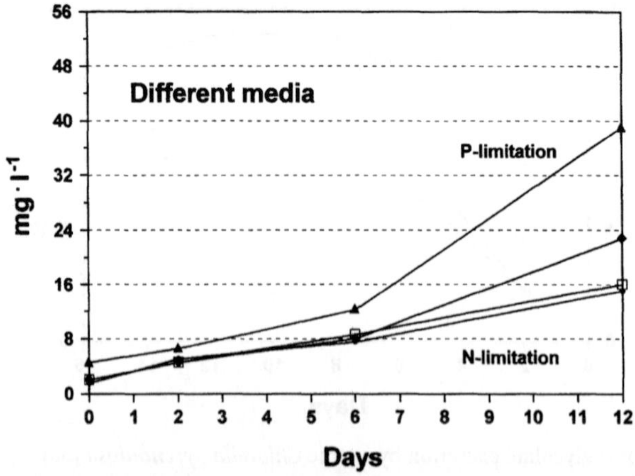

Fig. 3. Production of extracellular polysaccharides in cultures of *Chaetoceros affinis* in media with different nutrient limitation [36]

media with high N/P ratios, a more comprehensive study [39] using medium with 6 different N/P ratios was undertaken with the same diatom and in comparison with *Skeletonema costatum. S. costatum* produced only very limited amounts of extracellular polysaccharides but large amounts of cellular reserve polysaccharide. *C. affinis* again produced more extracellular polysaccharide at higher N/P ratios. This production was studied further in a separate experiment where *C. affinis* was cultured in a medium of 104 µM nitrate and 0.95 µM phosphate (N/P = 109). During the experiment nitrate was kept at a high concentration. Figure 4 provides cell counts and the concentrations of protein, cellular carbohydrate and extracellular polysaccharide. The exponential growth phase lasted for 6 days, but all the phosphate had been used up after 4 days. Net synthesis of protein stopped after 8 days and the net production of cellular carbohydrate after 16 days. The most striking feature of this culture was, however, that the production of extracellular polysaccharide proceeded at a relatively high rate after cell division had ceased, and net protein and cellular carbohydrate synthesis had stopped. This must mean that synthesis of the extracellular polysaccharide was the major photosynthetic activity under these conditions as synthesis of lipids is probably not large.

Obernosterer and Herndl [40] carried out a study in batch cultures of *C. affinis* with different N/P ratios in the medium and measured photosynthetic production and several other characteristics including extracellular release of organic material. The N and P composition of the media and specific productivity, photosynthetic extracellular release and percentage release of total primary production are shown in Table 4. During exponential growth the culture with N/P ratio 100 showed a PER of 29 compared to 15 for N/P 16, which is the Redfield ratio. Also the N-limited culture (N/P 5) was higher in PER than the N/P 16 culture. During stationary growth, the tendency of PER towards the N/P ratio is the same, but less pronounced. N/P 100 with highest PER coincided with

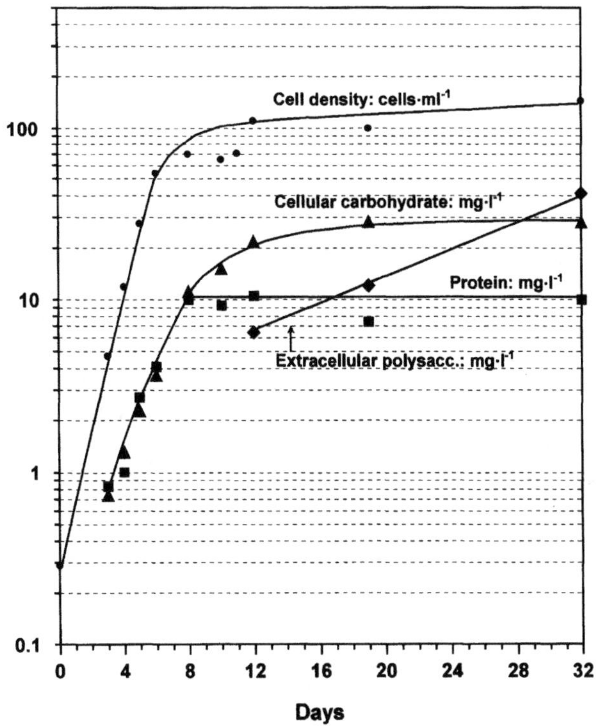

Fig. 4. Cell density.10^{-3}, production of cellular protein and carbohydrate and extracellular polysaccharide as a function of time for *Chaetoceros affinis* grown in a medium with a N/P ratio above 100 (see text for further explanation) [39]

Table 4. Specific photosynthetic production, extracellular release (ER) and PER as % of total primary production under different N/P ratios; means ± SE (n = 6) [40]

N:P ratio	Specific production µg C µg chl a^{-1} h^{-1}	ER µg C l^{-1} h^{-1}	PER
Exponential phase			
5	4.6 ± 1.9	7.0 ± 0.9	21 ± 9
16	4.4 ± 1.4	7.6 ± 2.5	15 ± 6
100	3.4 ± 1.1	10.2 ± 2.6	29 ± 10
Stationary phase			
5	1.0 ± 0.1	11.9 ± 3.6	30 ± 8
16	1.0 ± 0.4	12.3 ± 5.6	29 ± 8
100	1.2 ± 0.3	6.7 ± 1.8	37 ± 9

low absolute release. These authors used the same species, *C. affinis*, as Myklestad [39] in their study but inoculated the cultures with natural bacterial consortia and allowed them to grow under different N/P ratios. Because some of the released material may be taken up immediately by bacteria, they consider their PER values as conservative estimates. The main results of Obernosterer and Herndl [40] correspond with those of Myklestad and Haug [36] and Myklestad [39], who found the most marked increase in extracellular polysaccharides in cultures under P-limiting conditions.

3.2 Light and Temperature

Algae exposed to very high light often result in release of large proportions of photosynthate, possibly due to membrane damage. The percentage release is correlated to the relative inhibition of photosynthesis. The release is relatively unaffected by irradiance over an intermediate range [34]. Zlotnik and Dubinsky [31] found a tight coupling between rates of carbon fixation and of absolute rate of DOC excretion over a wide range of irradiances (0–1500 µquanta $m^{-2} s^{-1}$). In the range of irradiances from 10 up to 3000 µmol quanta $m^{-2} s^{-1}$ the percentage excretion was always below 10. At high irradiances this is in contrast to findings referred to in Hellebust [34]. Small portions of cellular carbon are lost to the medium during dark periods following periods of photoassimilation [30]. Mague et al. [22] stated that loss from the particulate fraction did not result in any significant increase in dissolved ^{14}C indicating respiration. The low excretion under these conditions may be due to lower concentrations of small metabolites in cells kept in the dark. It should be noted, though, that the same authors found a continuous release partly uncoupled from photosynthesis during short (2 h) light and dark periods. However, cells fixing $^{14}CO_2$ in the dark exhibited high excretion rates [31; 34].

There seem to be few studies carried out on the effect of temperature on release of DOC. Zlotnik and Dubinsky [31] used 2 freshwater species, *Chlorella vulgaris* and *Synechococcus* sp. and a marine, *Isochrysis galbana.* As mentioned above, *I. galbana* responded differently to temperature compared with the 2 other species. *C. vulgaris* and *Synechococcus* sp. showed highly sensitive absolute excretion rates. It is interesting to note a parallel temperature response of photosynthesis and excretion, although at the highest temperatures the PER rose to about 15 and 33, respectively. Absolute DOC excretion in *I. galbana* was independent of temperature, unlike photosynthesis. This is in accordance with results obtained by Verity [41] with the marine diatom *Leptocylindrus danicum.* This relative rate of excretion increases to about 25% at 5 °C and 50% at 35 °C. The authors think, and I agree that the pronounced increase in DOC percentage increase at the highest temperatures may result from a direct temperature effect on membrane transport [17].

3.3 Physiological Mechanisms

It does not seem that we have gained overwhelming new evidence on mechanisms for transport of excreted compounds across algal membranes during the last 10–20 years. We still think that small molecules like sugars, amino acids, organic acids and fatty acids mainly diffuse through the cell membrane [34]. Bjørnsen [42] made calculations of exudation based on assumptions that

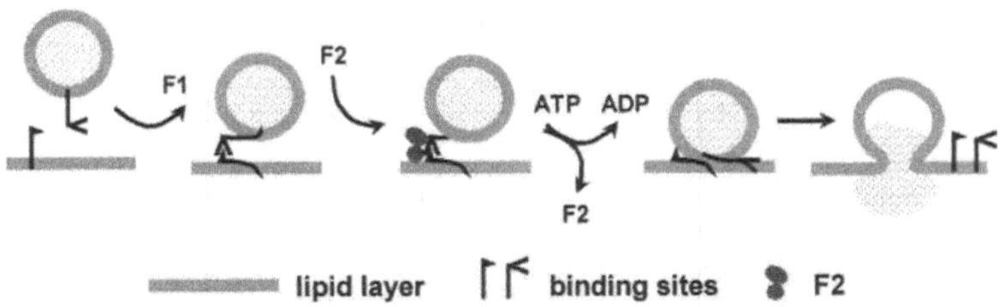

Fig. 5. A process of transport of biopolymers through a membrane mediated by transport vesicles with recognition, binding, and then fusing with the acceptor membrane to deliver the load [47]

given low molecular weight organic compounds which have a permeability of 10^{-9}cm · s^{-1} across cell membranes, the daily loss will be 50% of the intracellular pool when the transport is passive diffusion. The loss rate can be converted to 5% of carbon biomass per day. This model is interesting and would account for all of the observed release of small molecules from healthy phytoplankton, but his suggestion of continued exudation at night may be questioned as concentration of many small metabolites will decrease at night because of cellular dark reactions and would diminish the concentration difference across the membrane. The released DOC from phytoplankton, however, is often largely biopolymers [43–44]. Macromolecules such as proteins and polysaccharides are probably excreted by complex mechanisms, but the specific steps are largely unknown in algae.

Models and details are known from bacteria and human systems [45–47]. Proteins and polysaccharides are synthesized in the endoplasmic reticulum and Golgi bodies. Eukaryote cells have membrane-bound organelles to regulate the excretion and localization of macromolecules. Although each organelle has a distinct functional identity, membranes and proteins are continuously shuttled. This process is mediated by transport vesicles that bud from donor membranes (Fig. 5). Figure 5 indicates that transport vesicles are being bound to acceptor membranes with the help of energy-requiring reactions. After fusion, the macromolecules are delivered to the other side of the membrane. Somewhat similar processes have been identified when new cell wall material of algae is excreted through the plasma membrane by exocytosis of Golgi-derived vesicles containing organic cell wall material [48].

3.4 Release from Leaky Cells and Inefficient Feeding

So far we have mostly been dealing with release of soluble organic material from healthy cells, i.e. rapidly growing cells, although not necessarily growing at their maximum specific growth rate. It should be noted, however, that neither in culture nor in nature can we be sure to find exclusively cells in excellent condition. What we call healthy cells is therefore a best approximation. Healthy cells may change into leaky or bad cells when exposed to stress of some kind such as prolonged nutrient depletion, so that cell integrity is severely threatened or lost.

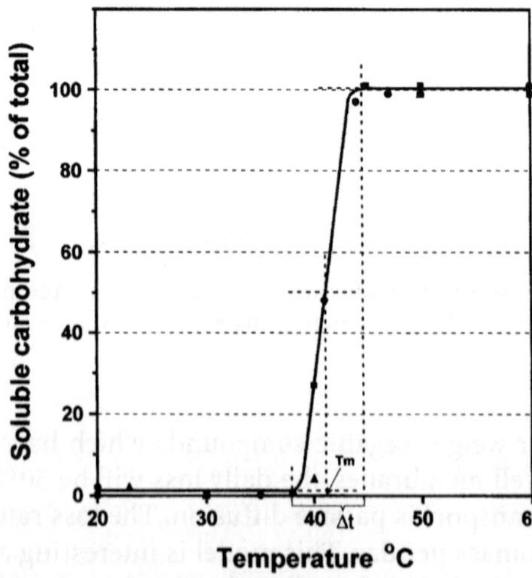

Fig. 6. Temperature effect on release of soluble carbohydrate for the marine diatom *Chaetoceros costatus*. T_m: transition temperature [17]

The soluble cell content will then freely move out of the cell and enter the DOC pool. This kind of release is not what we have defined as extracellular primary production. The phytoplankton plasma membrane is a very efficient barrier separating cell content from the outside water. If the membrane is broken it will undergo a dramatic increase in permeability, the integrity of the cell is destroyed and the cell dies. Figure 6 shows release of cell content of the marine diatom *Chaetoceros costatus* upon breaking its cell membrane by increasing the temperature. The T_m is an intrinsic property of the species and is defined as the temperature at which 50% of the cellular soluble material is found extracellularly at standard conditions. The shape of the curve indicates that the soluble cell content is being released over a narrow temperature range, which also reveals that the membrane structure is changing from being in an ordered to a disordered state [17].

How large a portion of carbon would leak out if the cell membrane is disrupted? In one case (N-limited culture) 40% was released; the non-dialyzable part of it contained principally carbohydrates (86%) and proteins (11%) [17]. Soluble cell content will vary according to growth phase, nutrient status and light conditions. The main fractions will be low molecular weight fractions, from 6–17% as measured with ^{14}C-incorporation and biochemical fractionation [49–50], soluble protein 10–20% of total cell protein [51], and soluble polysaccharides (glucans) 10% of cell carbon increasing to 60–80% under prolonged N-limitation in particular for diatoms [9]. Hama and Handa [52] measured the synthesis of carbohydrates by the ^{13}C-method in rapidly growing natural phytoplankton populations and reported 5–13% soluble carbohydrate-C of total cellular carbon. Wallen and Geen [53] found a ^{14}C-incorporation into the 80% ethanol-soluble fraction in surface phytoplankton from 16–68% of total incorporation; excretion was directly related to the size of this fraction.

Usually there is a large initial release of soluble organic material amounting to 20–50% of the organic carbon of the cells. Bratbak [54] made the point that the relative amount of soluble organic material vs insoluble structural macromolecules may vary from actively growing cells to senescence and death. Ittekkot [55] partly explained high concentrations of DOC towards the end of a phytoplankton bloom in the North Sea as release from physiologically old cells and release via cell lysis. Studies of autolysis kinetics in axenic culture of the marine diatom *Ditylum brightwellii* under nitrogen and phosphorus limitation revealed that nutrient stress was a significant cause of mortality in this species [56]. Van Boekel et al. [57] have provided a clear example from the field, where a *Phaeocystis* spring bloom declined after culmination caused by N-limitation, through cell lysis. The release of organic material was manifested by a steep increase of biomass of the microbial food web.

Phytoplankton may be exposed to other types of stress i.e. extreme light, change in salinity or pH, in addition to nutrient and temperature stress. In nature low light, as populations are sinking down the water column, nutrient stress, and biological factors such as virus and bacterial attack, may all be important as well. Milligan and Cosper [58] isolated a virus capable of lysing cultured cells of the chrysophyte *Aureococcus anophagefferens*. Radiotracer experiments with this species conducted by Gobler et al. [59] showed that 19% of the cell carbon was released to the dissolved phase by viral lysis. Surprisingly, apparently no nitrogen was lost from the cells.

Bratbak et al. [60] showed viral lysis of *Phaeocystis pouchetii* in cultures infected by the virus PpV01 recently brought into culture and shown to be lytic to *P. pouchetii* [61]. Viral infection disturbed the exponential growth and decimated the *Phaeocystis* population within 3 days. The concentration of DOC increased when the alga lysed. The produced DOC was approximately equal to the produced algal biomass, indicating that the entire algal biomass was converted to DOC upon cell lysis. This result differs significantly from the observation of Gobler et al. [59] in terms of the amount of organic material released.

Virus infection in *P. pouchetii* was further investigated in relation to cell growth and nutrition [62]. The alga was susceptible to virus infection in all stages of growth. Nutrient or light limitation of algal growth did not inhibit viral reproduction and cell lysis. However, the growth conditions of the host cells did have a significant impact on burst size; varying from a maximum of 510 viruses per host cell in exponentially growing cultures to a minimum of 15 in nutrient depleted or light limited cultures. There is growing evidence that virus infection is a quantitatively significant cause of mortality in a number of phytoplankton populations [63–64].

There are reports of aquatic bacteria that cause algal cells to lyse. Bacteria that lyse blue-green algae have been studied [65]. A marine gliding bacterium *Cytophaga sp.*, first shown to kill the noxious red tide flagellate *Chattonella antiqua* (Rapidophyceae), preys upon various species of marine phytoplankton. Imai et al. [66] studied the interaction between this bacterium and phytoplankton algae and found that all of the 5 rapidophycean flagellates, all of the four diatoms and one of 2 dinoflagellates were killed in some way by the cell wall being degraded by the bacterium (some of the algae did not have a real cell wall). The algae died usually a few days after inoculation of the bacteria to the algal culture. The authors express the view that bacteria such as *Cytophaga sp.* may be a significant factor influencing the population dynamics of phytoplankton in nature and may contribute to the sudden disappearance of red tides in coastal areas.

Entirely different is the reported killing of marine diatoms by the common bacteriovore flagellate *Cafeteria roenbergensis*. Nygaard and Hessen [67] observed that the flagellate attached to the surface of *Skeletonema costatum* cells 24 hours after inoculation of the diatom culture with the flagellate. After an aggregation step and decay of aggregates with involvement of bacteria, a complete dieback of the alga was the case after 7 days. The authors regarded this mechanism as being important in promoting termination of diatom blooms. Clearly, if this type of cell death is taking place in nature, a large part of the cell content will leak out when cells are disrupted or lysed.

Another kind of loss to the external organic pool is known to occur as a result of inefficient grazing or sloppy feeding. Laboratory grazing experiments have shown that there will be a cell breakage during feeding, with organic material leaking out or being suspended external to the feeding animal [68–70]. These authors found losses of from 4–35% of the ingested organic material, about half of this being soluble compounds. This is another example of cell breakage with the possibility of coincident transfer of organics to the external medium. This kind of DOC production will of course be a significant part of the total as grazing by zooplankton is a process taking place more or less intensively, all the time during the growing season.

3.5
Degradation and Solubilization of Insoluble Cellular Organic Material from Phytoplankton

When phytoplankton cells die for some reason (e.g. lysis) the soluble components will leave the cell to the external medium. The insoluble cellular material left has a relatively high chemical stability, but degradation may be greatly enhanced by enzymatically catalyzed reactions. Where are enzymes located? It seems less probable to find active enzymes suspended in sea water – most of them are extracellular enzymes located on the bacterial cell surface [71]. Active bacteria are partly free living in the water and partly attached to phytoplankton cells [71–73].

Those attached to cells are most important in degrading and dissolving the cell wall, debris from membranes, cell organelles, etc. The cell debris is composed of a conglomerate of macromolecules consisting of proteins, glycoproteins, polysaccharides, lipids, glycolipids and nucleic acids as the main components. Enzymes working on this material may partly be endoenzymes attacking someplace within a molecule to make it shorter and more soluble; and completely soluble at some point when, for example, no crosslinks are left between macromolecules in the network (Fig. 7). The catalyzed reactions are relatively simple – namely the hydrolysis of glycosidic, phosphodiester and peptide bonds in polysaccharides, nucleic acids and proteins, respectively. In

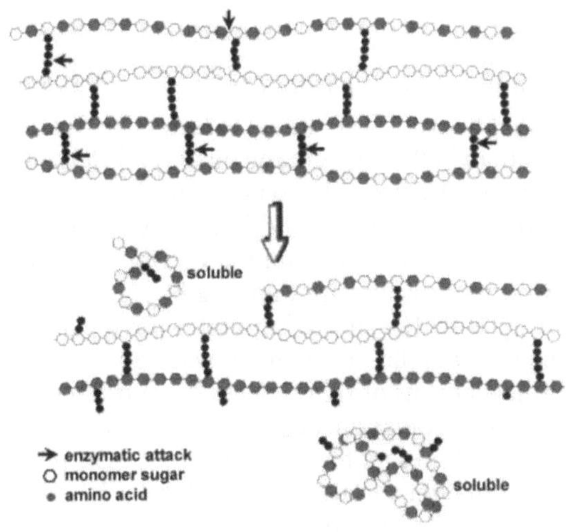

Fig. 7. Complex macromolecular network is being partly solubilized by enzymatic attack resulting in two soluble macromolecules being formed as indicated in the illustration

order to make food available for a bacterium, the biopolymer must be degraded further to monomers. Many enzymes have specialized on this task. These are called exo-enzymes which are splitting off one monomer at the time from one end of the molecule according to the specificity of the enzyme; as an example, a protease which is working from the amino-end of the molecule is eliminating one amino acid at a time from the amino-end of the protein molecule. From investigations of phytoplankton in culture and in the sea, significant observations have been made that bacteria solubilize much more organic material than they may be able to consume and therefore the remainder is being released as DOC.

Different mechanisms for this release have been proposed:

1. Bacteria are invading cells which are not growing and are degrading cellular material, which is being solubilized. During this process aggregation often take place and cells will finally die [74].
2. Bacteria are working on living cells and are peeling off polysaccharides and other biopolymers from the cell surface while the cell is still alive [75].
3. Specialized bacteria are attacking the alga and will kill living cells in a relatively short period of time i.e. 1/2 hour to a few days [65–66].

This close connection of the bacterium to the organic material will give rise to surface-associated hydrolysis to make food for uptake and the release of dissolved organic carbon to the external environment. Wangersky [76] made the point that in the ocean, nutrient regeneration is most efficient where surfaces exist. Release by enzymatic degradation on phytoplankton detritus may represent an important mechanism of DOC production [71]. If bacteria were able to utilize all the available soluble material from hydrolyzed organic material, nothing would be left to the DOC pool. There is evidence that this is not the case [77–78]. The rate of this type of release will obviously be difficult to estimate directly; the same will be true if one wants to know the fraction of the total release. Usually more than half of the organic material is left after lysis of phytoplankton cells. Theoretically the rest should be available for bacterial degradation when not grazed. On the other hand it has been shown that the turnover time for different fractions of phytoplankton detrital material may vary enormously [78–80] so if we basically are thinking of degradation in the euphotic zone we may end up with 10–40% of the total phytoplankton biomass. The work of Norrman et al. [81] illustrates that it was very difficult to discriminate between different mechanisms of DOC release i.e. extracellular production, cell lysis and heterotrophic degradation, in their mesocosm experiment. As referred to above, Smith et al. [75] present the mechanism of DOC production as an enzymatic hydrolysis of diatom surface macromolecules by the bacteria attached to living phytoplankton cells. The attached bacteria had extremely high specific growth rates and extracellular hydrolytic activities in their mesocosm experiment (see also Alldredge et al., [82]) and the writers think that the bacteria produce significant amounts of DOC from phytoplankton in this way in addition to the part being taken up and metabolized by the bacteria.

3.6 Field Investigations

The largest number of investigations on extracellular release have been carried out in cultures. During the last four decades, however, many field studies have been published, as reviewed by Hellebust [34], Wangersky [33], Fogg [83], Lancelot and Billen [84], Williams [85] and Münster [86]. When working in the field you are faced with the same problems as in laboratory studies, problems like contamination, sensitivity and accuracy of methods, only tiny populations, bacterial degradation, many pitfalls in using the principal ^{14}C-method, difficulties in separating the environmental variables light, temperature and nutrients. The taxonomic composition of phytoplankton populations are often unknown, and the difficulty in obtaining biomass measurements makes the measurement of DOC in the ocean a demanding task.

Lancelot [44] made kinetic experiments of 4-h duration; sometimes a lag in extracellular release and often a decrease in rate during the period (nonlinearity) which was attributed to heterotrophic bacterial uptake of extracellular products was apparent. Underestimates of phytoplankton extracellular release were considered important and may amount to 68%. Sometimes the kinetic curve was linear, indicating insignificant bacterial activity. The choice of a uniquely appropriate incubation time is thus difficult and kinetics will give a better approximation of phytoplankton release rates.

The magnitude of measured extracellular release from a number of investigations carried out from 1970 to 1998 with the ^{14}C-technique in waters ranging from eutrophic to oligotrophic is shown in Table 5 [22; 27; 44; 87–90; 93; 98–103]. Average values of PER range from 4–33% and gross variation from

Table 5. Extracellular release of soluble organic carbon as percentage of total photosynthetic production (PER) for natural phytoplankton populations from different areas.

Author	Year	Area	PER		Comment
			Range	Average	
Anderson & Zeutschel	1970	North-East Pacific Ocean	1–49	17	Eutrophic and oligotrophic w.
Thomas	1971	South-East coast of USA	0–44	21	Oligotrophic to entrophic areas
Choi	1972	W. North Atlantic Ocean	3–69	28	Grand banks; Scotian shelf; Gulf stream
Berman & Holm-Hansen	1974	Gulf of California	6–12	–	Eutrophic w.
			17–27	–	Oligotrophic w.
Williams & Yentsch	1976	Coast of Bahamas	0–23	7	Subtrophical Sea
Smith et al.	1977	Off coast North-West Africa	1–26	9	Upwelling area
Lancelot	1979	Southern North Sea, Europe	0–62	25	Belgian coastal w. and English Channel
Mague et al.	1980	Gulf of Maine, USA	5–30	–	Coastal w.
Wolter	1982	Inner Kiel Fjord, Germany	5–40	–	Dominant: *Skeletonema costatum*; *Prorocentrum micans*; Ch. sp.; nanoflagellates
Larson & Hagstrøm	1982	Baltic Sea, Sweden	4–22	14	Eutrophic w. and open regions
Jensen	1983	Randers Fjord, Denmark	10–28	15	Dominant: Diatoms and green flagellates
Lancelot	1983	Southern North Sea, Europe	4–80	33	English Channel
Lignell	1993	Tvärminne, Finland	–	4	Spring bloom
Karl et al.	1998	Station Aloha, Hawaii	25–34[a]	28	Vertical profile 5–100 m, June 1997
			12–15[b]	14	

[a] Nuclepore filters.
[b] Whatman-GF/F filters.

0–80%. Generally these values are higher than PER values from cultures in the exponential phase (Table 2), while stationary phase cultures may show higher PER than the actual field values shown in Table 5. I assume that the varying physiological state of phytoplankton populations in the field is one of the important factors in determining the relative release rate (PER).

The effect of irradiance on excretion was investigated by Smith et al. [88] but no changes in extracellular production were detected. Also Williams and Yentsch [87], Lancelot [89] and Karl et al. [90] apparently came to the same conclusion. This point thus seems to be fairly well established. The exception may be situations where surface photoinhibition takes place. In these cases the relative release may increase as a result of a reduced photosynthetic rate. There is convincing evidence, however, that phytoplankton algae exposed to direct sunlight have very high excretion rates [30].

Nutrient limitation in laboratory cultures has been shown to enhance the accumulation of extracellular products in the medium [18; 23; 37; 91–92; 113]. Some works in the field have failed to show an effect of nutrient limitation [89; 93] because this limitation is generally extremely difficult to demonstrate. Some of the more interesting contributions indicating release processes related to the availability of nutrients are some reports from the northern North Sea [94–95]. A phytoplankton bloom dominated by diatoms was investigated during a three month period (mid March to mid June). Distinct variation in concentrations of dissolved carbohydrates was found: 40–400 $\mu g\ l^{-1}$. Increased amounts of carbohydrates appeared in the water column simultaneously with a decrease in dissolved nutrients, notably inorganic nitrogen, whose concentration was below 1 µM during this period. Thus low nutrient concentrations and the resulting stagnation in phytoplankton growth probably caused the accumulation of dissolved carbohydrates. They measured glucose and galactose monomers in the water and considered them representative of the "storage" and "structural" polysaccharides after in situ degradation. An alternative or addition to their second suggestion could be extracellular polysaccharides since *Chaetoceros* sp. was the prominent component in the bloom. Smestad et al. [25] studied the chemical structure of the extracellular polysaccharide from the marine diatom *C. affinis* and found the monomer composition galactose:fucose:rhamnose to be about 1:1:1. In the North Sea bloom rhamnose and fucose were detected in addition to galactose which was analyzed all through the bloom. Lancelot [89] found no correlation between PER and inorganic nitrogen for diatoms. Eight observations ranged from 4–56 µM inorganic N and only one point was close to zero, which could be an indication of N-limitation – the other points certainly are not. These observations, therefore, may not be said to be in contradiction to the referred reports from the North Sea (see above). High PER, on the other hand, was observed related to N-limitation for phytoplankton populations dominated by dinoflagellates in the oligotrophic English Channel (46%) as well as for population dominated by flagellates (*Phaeocystis pouchetii)* in the Southern Bight, Belgium (70–80%).

Recently Karl et al. [90], as a part of a comprehensive study of the productivity at the Hawaii ocean station ALOHA, measured primary production including extracellular release as ^{14}C-DOC. Most interestingly they presented large and increasingly important ^{14}C-DOC fluxes and believe this to be a direct consequence of a major ecosystem change during the past decade, resulting in a shift from a primarily N-limited to a primarily P-limited habitat [96–97]. From observations over the last few years at station ALOHA an enigmatic accumulation of DOC in the euphotic zone has appeared. Earlier culture studies [36; 39; 40] indicated increased extracellular production as the N : P ratio increases. As shown in Fig. 4 [39] the production of extracellular polysaccharide may be the main photosynthetic activity under extreme P-limitation for this diatom. More recently in work with the same species, Obernosterer and Herndl [40] confirmed the basic earlier results (Table 4) and also suggested that bacterial activity may be suppressed under extreme P-limitation, leading to accumulation of dissolved organic material.

We may still expect the relationship between extracellular release and nutrient limitation in natural phytoplankton to be an area of conflicting opinions [85], although evidence for a positive correlation is increasing.

It has been shown by Wolter [93] in the Kiel Fjord that PER depends on the taxonomic composition of phytoplankton populations. *Chaetoceros* sp. released high amounts (27%), dinoflagellates low amounts, except *Prorocentrum micans* (20%), and the diatom *S. costatum* 5–12%. In this work it was possible to compare the different algal groups directly because nutrient concentrations were apparently never low and the physiological state of the phytoplankton was relatively stable. Hellebust [30] reported high excretion from *Chaetoceros* species (20%) whereas *Exuviella* sp. released only 3%. Myklestad [9] compared 7 marine diatoms in batch culture 12 days after exponential phase. Fig. 1 shows extracellular polysaccharide as percentage of the sum of cellular and extracellular polysaccharide for each alga, with ranges from 1% for *S. costatum* to 56% for *C. affinis*. Calculating extracellular polysaccharide as percentage of approximately total net photosynthetic production, the three *Chaetoceros* species range from 13- to 42%, the two *Thalassiosira species* release 4% and *S. costatum* 1%. If we estimate the release of small molecules as 5% then the range in total extracellular release would be from 6- to 48% among seven diatoms. These examples show clearly that the composition of the algal population has to be taken into consideration.

The relation between extracellular release and type of habitat – inshore-, coastal- or oceanic waters has not been settled [83; 85]. This range in habitat also implies decreasing supply of nutrients from eutrophic to oligotrophic situations. Part of the problem may be connected with differences in population density. Fogg argues that if a cell suspension is diluted, free diffusion of the metabolites concerned between the intra- and extracellular pools is being enhanced. This is consistent with results from culture work showing that the absolute amount of release per cell was found to rise as cell concentration was decreased [92]. Williams [85] does not think these results are convincing and presents contrasting data with no progressive change in PER of exponentially growing cultures of three different algal species. The increase in biomass was 50–100 fold [87]. He also found an average exudation of as low as 7% in "blue water" populations with very low chlorophyll concentrations. Smith and Wiebe [105] also concluded that the DOC release was directly proportional to biomass.

Recently Karl et al. [90], already mentioned above, in a substantial and carefully executed work on the productivity of the oligotrophic North Pacific Ocean, found relatively high release of organic matter – 25–34% (Table 5). When using Whatman GF/F glass fiber filters they estimated much lower PER values (12–15%). This was explained by adsorption of ^{14}C-DOC by the GF/F filters in contrast to very low adsorption by the nuclepore filters used. The authors suggest that the increasing importance of DOC-fluxes are a direct consequence of ecosystem changes and believe that the accumulation of photosynthetically derived organic matter in the surface waters is an indication of P limitation at Station ALOHA.

4
Composition and Chemical Nature of Released Compounds

4.1
Carbohydrates

Carbohydrate is the principal candidate to be the predominant extracellular product from phytoplankton both from direct [26; 33; 84; 106–107] and indirect evidence [90; 104; 108–110]. Wood and van Valen [108] pointed out that N- and P-lacking compounds appear to be relatively more common than N-containing compounds among the extracellular products, the principal exception being N-materials released from N-fixing cyanophyta. Baines and Pace [109] using their regression model based on 225 published observations, concluded that their data were not consistent with current models of passive diffusion in extracellular release. This means that the release could be more related to large molecule exudation, implying high molecular weight carbohydrates.

Table 6 shows an extended list of publications reporting the production of extracellular carbohydrate; some originating from natural populations [94–95; 111–112] and the rest are culture work. In most of these situations it is explicitly said that carbohydrate is the dominant part of the release products and as a rule the ratio monosaccharide (oligosaccharides) to polysaccharides is lower than 0.5. This means that polysaccharides probably constitute the largest fraction of all extracellularly produced components. Seven algal groups are represented counting members of diatoms, rhodophyceae, prymnesiophyceae, prasinophyceae, chlorophyceae, chrysophyceae and blue-green algae. Most of these polysaccharides are readily soluble in sea water but some of them are capsular polysaccharides [113–115] which means that the extracellular polysaccharide often forms a loose gel outside the cell wall and is partly soluble in sea water and often soluble in distilled water. Most of the large carbohydrate molecules are complex heteropolysaccharides with many component sugars and a complicated chemical structure, although simple glucans have been reported [95; 116–117]. It is not clear, however, if these glucans are released by leakage from unhealthy cells.

Many of the polysaccharides make viscous solutions; there are examples from several classes [118], the Rodophyceae [113; 119] and from diatoms [24]. Some diatom genera such as *Chaetoceros* seem to be excellent producers of ex-

Table 6. Extracellular release of carbohydrate

Author	Year	Carbohydrate total release	Monos. Polys.	Comment
Huntsman [121]	1972	–	0.1–0.4	*Dunallella tertioleeta*
Myklestad and Haug [36]	1972	>0.9	–	*C. affinis*
Myklestad et al. [24]	1972	>0.9	–	*C. affinis*, 13 l culture
Ramus [113]	1972	P	–	Capsular polysaccharide *Porphyridium aerugineum*
Ignatiades; and Fogg [92]	1973	–	–	*S. costatum*
Myklestad [9]	1974	–	–	4 out of 7 diatoms released copious amounts of poly-saccharides
Hellebust [34]	1974			Review
Ramus and Robins [122]	1975	P	–	*Porphyridium aerugineum*
Solter and Gibor [123]	1977	–	–	*Chlamydomonas Reinhardi*, mating process
Myklestad [39]	1977	95	–	*C. affinis*, various N/P ratios
Wangersky [33]	1978	P	–	Review
Ittekkot [94]	1981	P	low	Phytoplankton bloom
Ittekkot [95]	1982	P	low	Phytoplankton bloom
Lancelot [111]	1984	–	0.1–0.5	Polymers dominating
Jensen [107]	1984	P	–	Review article
Vieira and Myklestad [114]	1986	P	0.13–0.19	Ankistrodesmus densus
Hama and Handa [112]	1987	0.44	–	Lake water
Claus [133]	1988	P	0.23–0.5	Two marine diatoms
Marlowe [23]	1989	–	–	Carbohydrate present 0.8<1800D
Myklestad et al. [26]	1989	95	<0.2	*C. affinis*
Kroen and Ramus [119]	1990	P	–	*Rhodella reticulata*
Myklestad [115]	1991	P	low	*Prasinococcus capsulatus*
Goldman et al. [106]	1992	P	–	Large diatoms
Plude et al. [124]	1991	P	–	Cyanobacterium
Allard et al. [128]	1993	P	–	2 Chlamyclomonas species
Obernosterer and Herndl [40]	1995	P	0.4–0.5	*C. affinis* various N/P ratios
Mopper et al. [117]	1995	P	–	Mesocosm experiment
Myklestad [120]	1995	–	–	Review article
Borgaas [125]	1997	P	0.5	*Chrysochromulina polylepis*
Vieira et al. [126]	1908	–	–	*Synura petersenii* (Chrysophyceae)

P – carbohydrate predominating.

tracellular polysaccharides compared to, for example, *Thalassiosira* species [120]. The view that large carbohydrate production under conditions of negligible nutrient uptake and almost no cell division may lead to large discrepancies in estimates of new production has been put forward by Goldman et al. [106].

Is there any real evidence that what we measure as release of polymeric carbohydrate is extracellular polysaccharides and not just degradation or leakage of cellular products?

The comparison of the monosaccharide composition of extracellular polysaccharides and the corresponding cell wall polysaccharides for 3 marine diatoms shown in Table 7 reveals marked differences as indicated by the fucose/rhamnose ratio. Similar observations were made by Percival et al. [116]. Further, as shown by Myklestad et al. [24], the soluble cellular storage polysaccharide is a glucan and therefore completely different from the complex released polysaccharides. The glucan would also easily be detected and be an excellent indicator of leakage. It seems thus safe to conclude that extracellular polysaccharides apparently are unique molecules.

Table 8 [25; 114–116; 124–125; 127–129; 131–132; 171] provides monosaccharide and uronic acid composition and content of polyanionic half-ester sulfates of a number of reported extracellular polysaccharides. None of these are simple homopolymers but rather indicate really complex heteropolymers containing not less than 3 monosaccharide components. Many of them are charged, containing either sulfate groups or uronic acids (carboxyl groups). The polysaccharide produced by the blue-green *Microcystis flos-aquae* is composed mostly of galacturonic acid residues very like pectin and different from all the other extracellular polysaccharides compiled in Table 8. As far as reported none are linear and a number are heavily branched polymers [25; 116; 127–128]. The methylation analysis of the polysaccharide from *Coscinodiscus nobilis* (Table 8) showed that the molecule has branching points and that fucose and rhamnose likely occur as branching units. All neutral monosaccharides occur as non-reducing end groups and the main chain-units are (1→3)-linked fucose, 1→2)-linked rhamnose, (1→6)-linked mannose. Further chemical treatment revealed unhydrolysed oligouronides, which could be highly branched. These findings show that the chemical structure of this extracellular polysaccharide is very complex heteropolymer containing many monomers, several major chain units and has branching points. Major differences between the extracellular polysaccharide from *C. nobilis* and those of other diatoms investigated are the absence of fucose from the *N. frustulum* polysaccharide [129] and mannose and glucuronic acid from the polysaccharides of the *Chaetoceros* species *C. affinis and C. curvisetus* [25; 127]. Further details about the chemical structure are also given by Myklestad [120].

Table 7. Comparison of ratios between the major monosaccharides in polysaccharide fractions in three marine diatoms [25, 127, 171]

		Gal	Rha	Fuc	Fuc/Rha
Chaetoceros affinis	Extracellular polysaccharide	1	1.4	1.5	1.1
	Cell wall polysaccharide	1	0.6	2.9	4.8
Chaetoceros curvisetus	Extracellular polysaccharide	1	0.3	3.5	11.7
	Cell wall polysaccharide	1	0.5	1.0	2.0
Chaetoceros decipiens	Extracellular polysaccharide	1	1.8	2.0	1.1
	Cell wall polysaccharide	1	3.6	0.6	0.2

Table 8. Monosaccharide composition (%) and content of polyanionic sulfate in extracellular polysaccharides and a mucilage sample from the Adriatic Sea

Species	Gal	Glu	Man	Rha	Fuc	Ara	Xyl	Gluc A	Gal A	X_1	X_2	$-SO_3Na$	Author
Chrysochromulina polylepis	12	37	24	–	5	9	13	–	–	–	–	–	Borgaas (1997)
Chlamydomonas angustae	12	49	3	–	–	6	3	28	–	–	–	–	Allard and Tazi (1993)
Cylindroteca fusiformis	17	20	14	16	–	16	18	–	–	–	–	–	De Angelis et al. (1993)
Prasinococcus capsulatus	42	30	–	–	–	7	8	6	8	–	–	–	Myklestad (1999)
Microcystis flos-aquae	2	2	5	–	–	–	3	–	83	–	–	–	Plude et al. (1991)
Ankistrodesmus densus	21	14	11	3	–	10	10	–	–	4	26	–	Vieira and Myklestad (1986)
Adriatic mucilages	58	11	9	–	–	–	8	–	–	14	–	–	Marchetti (1992)
Coscinodiscus nobilis	–	16	19	15	34	–	6	9	–	–	–	16.7	Percival et al. (1980)
Nitzhia frustulum	8	–	34	24	–	–	–	–	–	34	–	–	Allan etal. (1972)
Chaetoceros affinis	26	–	–	39	35	–	–	–	–	–	–	8.9	Smestad et al. (1974)
Chaetoceros curvisetus	23	–	–	8	69	–	–	–	–	–	–	6.9	Smestad et al. (1975)
Chaetoceros decipiens	18	3	5	35	36	–	3	–	–	–	–	10.0	Myklestad and Haug (1976)
Chaetoceros debilis	29	5	10	17	30	–	9	–	–	–	–	not det.	Myklestad and Haug (1976)

As mentioned above, monosaccharides probably constitute a minor part when compared to total carbohydrate extracellular release. In culture work with *S. costatum* extracellular release of monosaccharides was 1 and polysaccharides 2.6 PER (Taraldsvik and Myklestad, to be published). Thus monosaccharides constitute 36% of the total in this case where measurements were made in the exponential phase of growth.

Sugar alcohols and organic acids have also been reported as extracellular products. Fogg [83] states that glycolic acid is often a major component of the released material, while Hellebust [34] found that out of 23 species of marine phytoplankton in only four did glycolate constitute more than 10% of total extracellular production. Recently Leboulanger et al. [130] reported measurement of glycolic acid by HPLC in the eastern tropical Atlantic Ocean. They found high values during the day, which then decreased during the night, and suggest that glycolate excretion by phytoplankton may be important. Since HPLC equipment is available in many laboratories, one would expect more reports on this rather conflicting matter.

4.2 Protein and Amino Acids

The C/N ratio according to Redfield is 6.6 for rapidly growing phytoplankton but much higher values are usually obtained for the extracellular production. This means a low content of N.

Peptides and amino acids are very common as release products of algae, but usually representing only a small fraction of the total extracellular production. An important exception are the blue-green algae, which often liberate very large portions of their photoassimilated N-substances into the medium (10–60%), [34]. Bratbak [54] found that *S. costatum* in culture excreted about half as much protein plus amino acids as total carbohydrate (Fig. 8). This alga is known to release small amounts of carbohydrate [9]. Claus [133], culturing the same diatom, found no protein release during the exponential phase and less than 10% of the extracellular carbohydrate in later phases. Another marine diatom, *Chaetoceros radicans,* released significant amounts of protein during the exponential phase as well. After extensive dialysis of extracellular products from the marine diatom *C. affinis* the protein constituted 1.5%, the rest being polysaccharide (see above) [24].

Extracellular enzymes from algae have been detected, for ex. phosphatases [134–135]. The earlier results on release of amino acids are summarized by Hellebust [34], showing liberation of amino acids from several algal classes. As methods of determination of amino acids have improved a number of publications have appeared in the literature since then.

The method of Lindroth and Mopper [136] and other HPLC-methods have been important in this move. Table 9 [22; 26; 137–143] shows an overview of some papers concerning excretion of amino acids from phytoplankton of which two are from outdoor tank experiments [137–138]. The main liberation of amino acids in the last-mentioned tank experiment occurred during daylight phases of the logarithmic growth and superimposed short-term

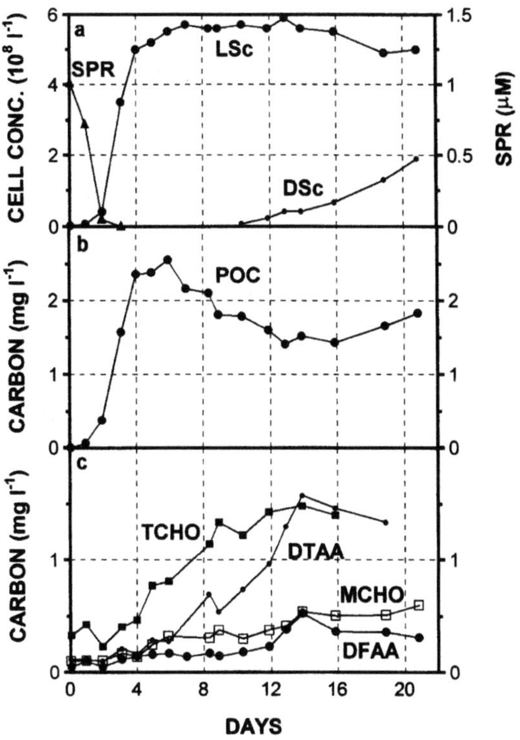

Fig. 8. Batch culture of axenic *Skeletonema costatum.* **a** Numbers of living *(LSc)* and dead *(DSc)* cells, and concentration of soluble P (reactive) *(SPR)*. **b** Concentration of particulate organic C *(POC)*. **c** Concentration of dissolved free *(DFAA)* and dissolved total *(DTAA)* amino acids (free + hydrolyzed), and of dissolved monosaccharides *(MCHO)* and dissolved total carbohydrates *(TCHO)* [54]

Table 9. Extracellular release of amino acids

Author	Year	Comment
Brockmann et al. [137]	1979	Outdoor tank
Mague et al. [22]	1980	*S. costatum*, intra and extracellular
Hammer and Brockmann [138]	1983	Outdoor tank
Poulet and Martin-Jezequel [143]	1983	*Chaetoceros debilis*
Admiraal et al. [139]	1984	Benthic diatoms
Haberstroh and Ahmed [140]	1986	*S. costatum*, N-deficient
Martin-Jesequel et al. [141]	1988	11 species, intra and extracellular
Myklestad et al. [26]	1989	*C. affinis*, different growth phases
Marsot et al. [142]	1991	Dialysis culture

variations showed a highly significant negative correlation with cell division activity.

This was possible to detect because the monoculture of *Thalassiosira rotula* employed showed partly synchronized cell divisions during exponential growth. Brockmann et al. [137] carried out combined measurements of dissolved amino acids and carbohydrates. Glucose and lysine occurred in highest concentrations. Mague et al., [22] found that extracellular production of free amino acids counted for 7.1% of the of the total extracellular C released in an exponentially growing culture of *S. costatum;* Myklestad et al., [26] measured 10.7% for *C. affinis* or 3.6% when calculated as percent of total incorporated cell N. In contrast to this Admiraal et al., [139] found that none of three benthic diatoms released more than 0.1% of the cellular N as free amino acids and concluded that benthic diatoms may act as net consumers of amino acids. Several authors did measure both intracellular and extracellular concentrations of many amino acids [22; 140–142]. The clear difference in relative composition of intracellular and extracellular fractions as pointed out by the first mentioned of these authors, show that the released pool is not just a portion of the intact cells' content.

When all reports in Table 9 are viewed as a whole the following are the most prominent free amino acids extracellularly: Ser; Gly; Lys; Ala; Glu; Asp; Orn and His.

Unspecific excretion of organic N has also been reported. Release of organic N which may represent up to 75% of the inorganic N-uptake took place in the early part of the incubation in algal batch cultures. This was followed by organic N uptake when inorganic N was exhausted [144–146]. Under conditions of natural irradiance degradation of nitrogenous biomass during day was observed, with intense exchange of material between the particulate and dissolved N fractions [145]. The released pool of organic N-containing material has been suggested as not being amino acids and is as yet not identified.

4.3 Lipids and Fatty Acids

The information given in the literature on excretion of lipids and fatty acids is very limited. Hellebust [34] reported 2.8–10.7% of total extracellular release as chloroform extract of media of growing marine phytoplankton cultures. Billmire and Aaronson [147] measured extracellular lipid production from the freshwater phytoflagellate *Ochromonas danicus* and found excretions of 8% of the cellular dry weight. The lipid fraction was mainly composed of free fatty acids, sterols and sterol esters, pigments and further sulfolipids, glycolipids and phosphatides. Parrish and Wangersky [148] examined the cellular and extracellular production of lipid classes in cage culture turbidostats of the marine diatom *Phaeodactylum tricornutum* at different N-nutrition levels. These experiments are particularly interesting because they are carried out in continuous culture and designed especially for lipid and lipid class production.

A decrease in total lipids per cell and in total lipid production with increasing N-stress was found. However, there was a change in types of lipids produced as the alga made the transition from N-replete to N-deficient media. Different classes of dissolved lipids were produced under these two conditions of N-supply. It is interesting to note that the absolute production rate of dissolved lipid per cell was practically constant and not influenced by N-limitation and consistent with, for example, excretion of carbohydrate and amino acids from another marine diatom [26]. However, if the proportion of total lipid produced is calculated, the percentage increased from 10% at N-replete conditions to 25% under N-stress. These values are comparable to published PER values of phytoplankton (Table 2). From this work one should also note the observation not directly related to extracellular production, that cellular synthesis of storage lipids, triglycerides, was triggered by N-limitation, while synthesis of membrane-associated polar lipid classes was reduced under these conditions.

4.4 Vitamins

Carlucci and Bowes [29] showed that vitamin production in phytoplankton algae was attributed to release during exponential growth and upon cell death and lysis in old cultures. Vitamin utilization was readily observed in cultures of two species; *S. costatum* produced utilizable biotin for *Amphidinium carterae*. The amount of utilizable vitamin and the rate at which it was exuded depended on the algal species and conditions of culturing. Aaronson et al. [149] showed that when *O. danicus* (chrysophyceae) was grown on a defined medium the cells excreted a number of vitamins including riboflavin, vitamin E and nicotinic acid in addition to four amino acids. Swift [150] published an excellent review of phytoplankton production, excretion and utilization of vitamins.

4.5 Toxins

It is well known that some algal species produce substances toxic to fish and other organisms. The main part of this type of work is currently being published in the Proceedings from the International Conference on Toxic Marine Phytoplankton, of which eight have been held up to now. The extent to which the toxic algae also produce extracellular toxins is not well known. In this review a few cases will be mentioned.

The toxin production of the prymnesiophyte *Prymnesium parvum* has been known for 60 years; it has been shown that the alga excreted the toxin extracellularly [151–152]. The alga has caused fish deaths in brackish fish ponds and other brackish habitats and lately in a large marine fjord system [153].

The nature of the toxin has been investigated over a long period of time and the latest work has been reported by Igarashi et al. [154–155]. The chemical structure of one of two main components of the toxin has been elucidated and was called prymnesin 2. It is a glycoside with the molecular formula $C_{96}H_{136}Cl_3NO_{35}$. The molecule contains ether rings, conjugated double and triple bonds and an uncommon L-xylose.

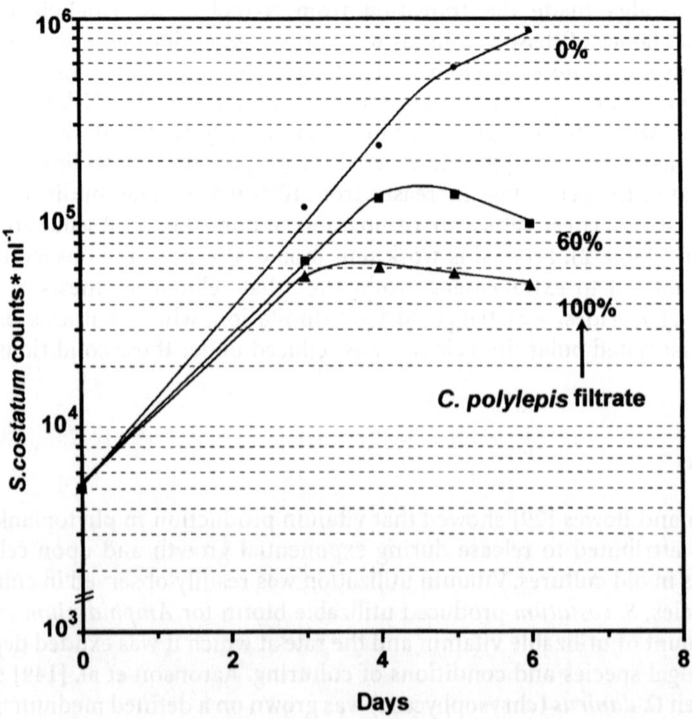

Fig. 9. Effect on growth of *Skeletonema costatum* by addition of increasing proportions of cell-free filtrates of *Chrysochromulina polylepis* culture (2.5×10^5 cells · ml^{-1}) [156]

The prymnesiophyte *Chrysochromulina polylepis,* known from an extensive bloom in the Kattegat and the Skagerrak 1988 (see below), releases toxic components into the medium in culture [156]. Figure 9 shows a *S. costatum culture* growing in medium with increasing addition of *C. polylepis* culture filtrate. The growth rate as well as final yield of *S. costatum* after six days were markedly affected. Isolated toxin fraction from *C. polylepis* culture filtrate tested by different test methods revealed significant activity [157]. Arzul et al. [158] showed that diatom growth was repressed during an offshore bloom of the dinoflagellate *Gyrodinium* cf. *Aureolum.* Water samples from high density blooms were shown to decrease the growth rate of *Chaetoceros gracilis* in bioassays.

The red tide of the dinoflagellate *Cochlodinium* "type 90" in China in 1990 caused high mortality of marine organisms, including fish. On the basis of the observed phenomena it was suggested that this organism released toxins into the water. The rapidophyte *Heterosigma akashiwo* has been shown to excrete allelopathic substances, reported to be a polysaccharide-protein complex, which suppresses the growth of *S. costatum* and enhances the growth of *Prorocentrum triestinum* [159]. This organism lacks a cell wall and has a single cell membrane. *Prorocentrum lima,* a benthic dinoflagellate found in temperate and tropical waters, is known to produce okadaic acid. Culture studies indicated that nearly 20% of the toxin was released from the growing cell to the external medium.

4.6 Other Substances

A long list of additional compounds being released could be given. Among these are volatile-, phenolic-, chemotactic-, growth inhibiting and growth promoting compounds.

5 Special Events

5.1 Adriatic Sea Mucilage

Large quantities of mucilaginous material occurred in the Northern Adriatic Sea during the summer months of 1988,1989 and 1991. It appears as different stages in particle size from transparent particles to macroflocs, clouds (from a few cm to 4–5 m long), creamy surface layers to surface gelatinous masses. Part of the material floating on the sea surface was deposited on beaches by currents and wind, threatening bathing and other tourist activities. Serious problems for the fisheries were created by suspended and sinking mucilaginous material. Pycnoclines are thought to be important in mucilage aggregation, retarding and collecting the sinking material. Although occurrence is irregular, mucilage incidents were recorded from the Adriatic Sea some 200 years ago. The intensity of the recent phenomenon has been reported only a few times earlier [160].

There seems to be a wide consensus that the mucilage represents a buildup of organic and inorganic material entrapped in a matrix, produced by gelling of extracellular polysaccharides released mainly by phytoplankton. Observations by divers indicate that gelling substances were produced in the upper part of the water column and that accumulation took place in the pycnocline layers. Among phytoplankton algae, diatoms are thought to be important as producers of polysaccharides (especially *Nitzschia closterium and C. affinis).*

Environmental conditions inducing large extracellular production of polysaccharides are the most likely causes of triggering the mucilage events. Experiments have indicated (see above) that a high N/P ratio in the water, with low nutrient concentration, is one of the conditions stimulating exudation. Field observations of long-lasting blooms of extracellular polysaccharide-producing diatoms in low nutrient (high N/P ratio) conditions have been reported.

The Northern Adriatic is strongly influenced by the Po river water which has a high N/P ratio (up to 60), thus partly explaining high N/P ratios measured in Northern Adriatic waters (Degobbis et al.; Azam – to be published; [39; 40; 160–161].

5.2 The *Chrysochromulina polylepis* Bloom 1988

During May and June 1988, an unusual bloom of the marine prymnesiophycean flagellate *C. polylepis* occurred in Scandinavian waters in the area of the

Kattegat and the Skagerrak and eventually covered an area of about 75,000 km^2. The bloom was first observed in the Northern part of the Kattegat; it spread along the Swedish and Norwegian coasts and to the inshore areas of the Kattegat, the Storebælt and the Øresund. The bloom was not obvious visually because maximal algal concentrations were often found at the pycnocline depth of 10–20 m. The highest concentrations, 100×10^6 cells ml^{-1} was recorded in the Danish Kattegat. At the beginning of the bloom, *C. polylepis* was found together with other phytoplankton species, but as the bloom developed further it changed completely to be dominated by *C. polylepis*. Severe toxic effects were observed in areas containing high concentrations of *C. polylepis*. The industry of salmon and trout farming suffered considerable losses of fish. Furthermore, the bloom had a marked effect on the littoral and sublittoral communities down to the pycnocline. Deleterious effects were observed on sessile plants (red and brown seaweeds), mollusks (bivalves and snails), and echinoderms (sea stars and sea urchins),while crustaceans (crabs) seemed less affected [162]. A low incorporation rate of thymidine in the pycnocline suggests that the low biomass measured was due to low bacterial production rather than grazing. This together with laboratory investigations, suggests that *C. polylepis* had a deleterious effect on natural populations of marine bacteria [163]. During the same investigation moribund cells of *Ceratium* sp. cells occurring in the *C. polylepis* layer were observed. Based upon these reports and the above-mentioned work on bialgal cultures and effects of culture filtrates and isolated toxin, there seems little doubt that growing *C. polylepis* excretes toxic substances to the environment. These toxins seem to act similarly to the toxins from the closely related species *Prymnesium parvum*. Nothing is seen to conflict with the interpretation that toxins exert their effect by disrupting cell membranes [164]. Graneli et al. [165] named this bloom "the most spectacular toxic algal bloom hitherto recorded"; it came as a total surprise.

5.3
Phaeocystis sp. Blooms

Foam accumulation may be observed for some days every spring on the beaches along the French, Belgian, Dutch and German continental coastal waters of the North Sea, causing economic as well as serious environmental problems. The biological reality of this is transient accumulations of mucilaginous aggregates, occurring under specific windy conditions. These events are some of the most spectacular consequences of the proliferation in this nutrient-rich coastal water of one single phytoplankton organism [166], the colony-forming *Phaeocystis* sp. In these waters *Phaeocystis* succeeds a silica-controlled diatom population. It dominates the spring phytoplankton with more than 95% of biomass and cell number and consists almost entirely of the colony form. Diatoms co-occurring with *Phaeocystis* are typically mucilage producing *Chaetoceros* sp. and *Rhizosolenia* sp. At the early beginning of the bloom, free-living cells are transformed into colonies, which increase in size and number during the exponential phase. At the culmination of the bloom the population is composed exclusively of large colonies of which mucilaginous matrix accounts for 90% of the colony biomass [167] although so large a fraction of polysaccharides has been questioned by van Rijssel et al., [168]. Low phosphate concentrations are shown to be one of the inducing factors for colony formation from single cells [169]. Dissolved extracellular organic substances as extracolonial production [170] and derived from the disruption of healthy and senescent colonies, transiently accumulate in the water during the later phases of the bloom. The polymeric structure of this mucilaginous material, with carboxylated and sulfated polysaccharide chains, makes it suitable for foaming under specific physical conditions of wind and currents.

6 Conclusions

- The existing disagreement in 1997 on the question "Do healthy phytoplankton excrete organic products?" has been settled and there is now convincing evidence that even the most healthy phytoplankton cells do release some of their photosynthetic production to the external medium as a normal process. This agreement has the consequence that primary production means cellular plus extracellular production; the importance of which has been clearly demonstrated recently.
- A number of more recent publications have clearly indicated that extracellular production is taking place during all growth phases. When high PER values were reported in older work this was easily misinterpreted as the excretion rate being highest in the stationary phase, because the photosynthetic production is generally low in this phase,. Therefore, as has been noted by several workers, the results from the measurement of excretion should be given as percent of total production (PER) and as a specific rate as for ex. Pg $cell^{-1}$ day^{-1}.
- It is well established that phytoplankton cells limited in N or P may continue the photosynthetic production of cellular and extracellular compounds not having the limiting nutrient as a constituent, often leading to high PER values. On the other hand, the specific exudation rate may not be higher than under nutrient replete conditions. High medium N/P ratio has been shown to increase the production of extracellular polysaccharides.
- There is now growing evidence that excretion is tightly coupled to carbon fixation over a wide range of irradiances and that PER is never high except at extreme irradiances comparable to full sunlight. The number of investigations of extracellular release as a function of temperature are low, but the finding that some species show an absolute excretion rate independent of temperature and a high PER at temperatures well above the optimal one, and other species exhibit a parallel response of photosynthesis and absolute release, is interesting to note.
- During rapid growth under natural conditions the extracellular production may amount to 5–10% which probably is the larger part of the release under such conditions. Inefficient feeding will always make a contribution, while cell lysis caused by nutrient or other environmental stress, virus or bacteria may be quantitatively more important during slow growth or stagnant

conditions. Extracellular bacterial attack should also be noted. Under these sub-optimal conditions extracellular release from 30–80% seems to be realistic.

- When looking at composition of released compounds, carbohydrate constitutes the major fraction according to evidence from a large number of reports. Most of this material seems to appear as polysaccharides. Sugar alcohols and organic acids often make a significant contribution. The next largest fraction may be N-compounds with the most prominent members being protein/polypeptides and amino acids. Lipids also make an important group of substances although much smaller than the other two main fractions. Some extracellular products although low in concentration may appear to be ecologically important, for ex. growth inhibiting/toxic compounds such as those released during the *Chrysochromulina polylepis* outbreak in 1988.
- The physiological mechanisms behind phytoplankton exudation are still poorly understood. Two more recent hypothesis should be mentioned: Bjørnsen [42] looks upon algal exudation as a process of passive diffusion through the cell membrane, while Wood and Van Valen [108] propose the process would be adaptive under a certain set of conditions where nutrient and light availability are temporarily uncoupled, when the capacity of the cell to accumulate storage molecules must be exceeded. A specific situation would be nutrient limitation, excess of light energy and with cells filled up with storage molecules.
- The importance of excretion, apart from being an integral part of the primary production, has been demonstrated here through three examples of special events involving extracellular release with a significant impact on the environment and human activity.

7 References

1. Toggweiler JR (1992) Nature 356:665
2. Bauer JE, Williams PM, Druffel ERM (1992) Nature 357:667
3. Kirchman DL, Suzuki Y, Garside C, Ducklow HW (1991) Nature 352:612
4. Carlson CA, Ducklow HW (1996) Aquat Microb Ecol 10:69
5. Zweifel UL, Norrman B, Hagstrom A (1993) Mar Ecol Prog Ser 101:23
6. Wangersky PJ (1993) Mar Chem 41:61
7. Hansell DA (1993) Mar Chem 41:195
8. Fitzwater SE, Martin JH (1993) Mar Chem 41:179
9. Myklestad SM (1974) J Exp Mar Biol Ecol 15:261
10. Fernandez-Reiriz MJ et al. (1989) Aquaculture 83:17
11. Parsons TR, Stephens K, Strickland JDH (1961) J Fish Bd Canada 18:1001
12. Brown MR, Jeffrey SW (1992) J Exp Mar Biol Ecol 161:91
13. Pugh PR (1975) Mar Biol 33:195
14. Fabregas J, Herrero C, Cabezas B,Abalde J (1986) Aquaculture 53:101
15. Werner D (1970) Helgol wiss Meeresunters 20:97
16. Dortch Q, Clayton JR, Thoresen SS, Ahmed SI (1984) Mar Biol 81:237
17. Myklestad SM, Swift E (1998) Eur J Phycol 33:333
18. Sharp JH (1977) Limnol Oceanogr 22:381
19. Fogg GE (1977) Limnol Oceanogr 22:576
20. Aaronson (1978) Limnol Oceanogr 23:838

21. Sharp JH (1978) Limnol Oceanogr 23:839
22. Mague TH, Friberg E, Hughes DJ, Morris I (1980) Limnol Oceanogr 25:262
23. Marlowe IT, Rogers LJ, Smith AJ (1989) Mar Biol 100:381
24. Myklestad S, Haug A, Larsen B (1972) J Exp Mar Biol Ecol 9:137
25. Smestad B, Haug A, Myklestad S (1974) Acta Chem Scand B28:662
26. Myklestad S, Holm-Hansen O, Vaarum KM, Volcani BE (1989) J Plankton Res 11:763
27. Berman T, Holm-Hansen O (1974) Mar Biol 28:305
28. Laws EA, Wong DCL (1978) J Phycol 14:406
29. Carluzzi AF, Bowes PM (1970) J Phycol 6:393
30. Hellebust JA (1965) Limnol Oceanogr 10:192
31. Zlotnik I, Dubinsky Z (1989) Limnol Oceanogr 34:831
32. Marker AFH (1965) J mar biol Ass UK 45:755
33. Wangersky PJ (1978) Production of dissolved organic matter. In: Kinne O (ed) Marine Ecology, Vol IV. Dynamics. J Wileys and Sons, New York, chap 4
34. Hellebust JA (1974) Extracellular products. In: Stewart WD (ed) Algal physiology and Biochemistry. Univ California, USA, 838
35. Nelson EB, Tobert NE (1969) Biochim Biophys Acta 184:263
36. Myklestad S, Haug A (1972) J Exp Mar Biol Ecol 9:125
37. Hellebust JA (1967) In: Lauff GH (ed) Estuaries. Am Assoc Adv Sci Publ 83:361
38. Nalewajko C (1978) Release of organic substances.In: Hellebust JA, Craigie JS (eds) Handbook of Phycological Methods, Physiological and Biochemical Methods. Cambridge University press, Cambridge, UK, 389
39. Myklestad S (1977) J Exp Mar Biol Ecol 29:161
40. Obernosterer I, Herndl GJ (1995) Mar Ecol Prog Ser 116:247
41. Verity PG (1981) J Exp Mar Biol Ecol 55:159
42. Bjørnsen PK (1988) Limnol Oceanogr 33:151
43. Lignell R (1990) Mar Ecol Prog Ser 68:85
44. Lancelot C (1979) Mar Ecol Prog Ser 1:179
45. Palade G (1975) Science 189:347
46. Verner K, Schatz G (1988) Science 241:1307
47. Bock JB, Scheller RH (1997) Nature 387:133
48. Van den Hoek C, Mann DG, Jahns HM (1995) Algae. An introduction to phycology. Cambridge Universityn Press. Cambridge, UK
49. Smith REH, Geider RJ (1985) J Exp. Mar Biol Ecol 93:191
50. Terry KL,Hirata J, Laws EA (1983) J Exp Mar Biol Ecol 68:209
51. Kirkvold S (1994) Thesis NTH, University of Trondheim
52. Hama J, Handa N (1992) J Exp Mar Biol Ecol 162:159
53. Wallen DG, Geen GH (1971) Mar Biol 10:157
54. Bratbak G (1988) Dr phil thesis, University of Bergen, Norway
55. Ittekkot V (1982) Mar Chem 11:143
56. Brussard CPD, Noordeloos AAM, Riegman R (1997) J Phycol 33:980
57. van Boekel WHM, Hansen FC, Riegman R, Bak RPM (1992) Mar Ecol Prog Ser 81:269
58. Milligan KLD, Cosper E (1994) Science 266:805
59. Gobler et al. (1997) Limnol Oceanogr 42:1492
60. Bratbak G, Jacobsen A, Heldal M (1998) Aquat Microb Ecol 16:11
61. Jacobsen A, Bratbak G, Heldal M (1996) J Phycol 32:923
62. Bratbak et al. (1998) Aquat Microb Ecol 16:1
63. Bratbak G, Egge JK, Heldal M (1993) Mar Ecol Prog Ser 93:39
64. Cottrell MT, Suttle CA (1995) Limnol Oceanogr 40:730
65. Daft MJ, Stewart WDP (1973) New Phytol 72:799
66. Imai I, Ishida Y, Hata Y (1993) Mar Biol 116:527
67. Nygaard K, Hessen DO (1994) Nature 367:520
68. Lampert W (1978) Limnol Oceanogr 23:831
69. Olsen Y, Vaarum KM, Jensen A (1986) J Plankton Res 8:505
70. Roy S, Harris RP, Poulet SA (1989) Mar Ecol Prog Ser 52:145

71. Azam F, Cho BC (1987) Bacterial utilization of organic matter in the sea. In: SGM 41, Ecology of microbial communities. Fletcher et al. (eds) Cambridge University Press, Cambridge, UK
72. Thingstad F, Billen G (1994) J Mar Syst 5:55
73. Middelboe M, Søndergaard M, Letarte Y, Borch NH (1995) Microb Ecol 29:231
74. Biddanda BA, Pomeroy LR (1988) Mar Ecol Prog Ser 42:79
75. Smith DC, Steward GF, Long RA, Azam F (1995) Deep-Sea Res II 42X:75
76. Wangersky PJ (1977) Helgol wiss Meeresunters 30:546
77. Jacobsen TR, Azam F (1985) Bull Mar Science 35:495
78. Biddanda BA (1988) Mar Ecol Prog Ser 42:89
79. Pett RJ (1989) Mar Ecol Prog Ser 52:123
80. Chen W, Wangersky PJ (1996) J Plankton Res 18:1521
81. Norrman B, Zweifel UL, Hopkinson CS, Fry B (1995) Limnol Oceanogr 40:898
82. Alldredge AL, Gotschalk C, Passow U, Riebesell U (1995) Deep-Sea Res II 42:9
83. Fogg GE (1983) Bot Marina 26:3
84. Lancelot C, Billen G (1985) Adv Aquat Microbiol 3:263
85. Williams PJleB (1990) Mar Microbial Food Webs 4:175. In: Maestrini SY, Rassoulzadegan F (eds) Microbial growth: inputs and losses,practical approaches
86. Münster U (1993) Antonie van Leeuwenhoek 63:243
87. Williams PJleB, Yentsch CS (1976) Mar Biol 35:31
88. Smith WO, Barber RT, Huntsman SA (1977) Deep-Sea Res 24:35
89. Lancelot C (1983) Mar Ecol Prog Ser 12:115
90. Karl DM, Hebel DV, Bjørkman K, Letelier RM (1998) Limnol Oceanogr 43:1270
91. Guillard RRL, Wangersky PJ (1958) Limnol Oceanogr 3:449
92. Ignatiades L, Fogg GE (1973) J mar biol Ass UK 53:937
93. Wolter K (1982) Mar Ecol Prog Ser 7:287
94. Ittekkot V, Brockmann U, Michaelis W, Degens ET (1981) Mar Ecol Prog Ser 4:299
95. Ittekkot V, Degens ET, Brockmann U (1982) Limnol Oceanogr 27:770
96. Karl et al. (1995) Nature 373:230
97. Karl et al. (1997) Nature 388:533
98. Anderson GC, Zeutschel RP (1970) Limnol Oceanogr 15:402
99. Thomas JP (1971) Mar Biol 11:311
100. Choi CI (1972) Deep-Sea Res 19:731
101. Larsson U, Hagstrøm Å (1982) Mar Biol 67:57
102. Jensen LM (1983) Mar Ecol Prog Ser 11:39
103. Lignell R et al. (1993) Mar Ecol Prog Ser 94:239
104. Børsheim et al. (1999) Mar Chem 63:225
105. Smith DF, Wiebe WJ (1976) Appl Environ Microbiol 32:75
106. Goldman JC, Hansell DA, Dennett MR (1992) Mar Ecol Prog Ser 88:257
107. Jensen A (1984) Excretion of organic carbon as function of nutrient stress. In: Holm-Hansen O, Bolis L, Gilles R (eds) Marine phytoplankton productivity, Springer, Berlin Heidelberg New York, p 61
108. Wood AM, Van Valen LM (1990) Mar Microbiol Food Webs 4:103. In: Maestrini SY, Rassoulzadegan F (eds) Microalgal growth: inputs and losses, practical approaches
109. Baines SB, Pace ML (1991) Limnol Oceanogr 36:1078
110. Williams PJleB (1995) Mar Chem 51:17
111. Lancelot C (1984) Estuar Coastal Shelf Sci 18:65
112. Hama T, Handa N (1987) Arch Hydrobiol 109:227
113. Ramus J (1972) J Phycol 8:97
114. Vieira AAH, Myklestad S (1986) J Plankton Res 8:985
115. Myklestad SM (1999) J Phycol 35:1032
116. Percival E, Rahman MA, Weigel H (1980) Phytochemistry 19:809
117. Mopper K et al. (1995) Deep-Sea Res II 42:47
118. Painter TJ (1983) Algal polysaccharides. In: Aspinal OG (ed) The polysaccharides vol 2. Academic Press, New York

119. Kroen WK, Ramus J (1990) J Phycol 26:266
120. Myklestad SM (1995) Sci of the Total Environ 165:155
121. Huntsman SA (1972) J Phycol 8:59
122. Ramus J, Robins DM (1975) J Phycol 11:70
123. Solter KM, Gibor A (1977) Plant Sci Lett 8:227
124. Plude JL et al. (1991) Appl Environ Microbiol 57:1696
125. Borgaas H (1997) Thesis. Norwegian University of Science and Technology
126. Vieira AAH, Lombardi AT, Sartori AL (1998) Phycologia 37:357
127. Smestad B, Haug A, Myklestad S (1975) Acta Chem Scand B 29:337
128. Allard A, Tazi A (1993) Phytochemistry 32:41
129. Allan GG, Lewin J, Johnson PG (1972) Botanica Mar 15:102
130. Leboulanger C et al. (1998) J Phycol 34:651
131. De Angelis F et al. (1993) Phytochemistry 34:393
132. Marchetti R (1992) In: Vollenweider RA, Marchetti R, Viviani R (eds) Marine Coastal Eutrophication. Elsevier,Amsterdam, p 21
133. Claus W (1988) Mitt Inst Allg Bot Hamburg 22:63
134. Aaronson S (1971) Limnol Oceanogr 16:1
135. Kuenzler EJ, Perras JP (1965) Biol Bull Mar Biol Lab Woods Hole 128:271
136. Lindroth P, Mopper K (1979) Anal Chem 51:1667
137. Brockmann UH et al. (1979) Mar Ecol Prog Ser 1:283
138. Hammer KD, Brockmann UH (1983) Mar Biol 74:305
139. Admiraal W, Laane RWPM, Peletier H (1984) Mar Ecol Prog Ser 15:303
140. Haberstroh PR, Ahmed SI (1986) J Exp Mar Biol Ecol 101:101
141. Martin-Jezequel V et al. (1988) Mar Ecol Prog Ser 44:303
142. Marsot P, Cembella AD,Colombo JC (1991) J Phycol 27:478
143. Poulet SA, Martin–Jezequel,V (1983) Mar Biol 77:93
144. Collos Y (1992) Mar Ecol Prog Ser 90:201
145. Collos Y, Døhler G, Biermann I (1992) J Plankton Res 14:1025
146. Collos Y et al. (1998) C R Acad Sci Paris, Life Sciences 321:673
147. Billmire E, Aaronson S (1976) Limnol Oceanogr 21:138
148. Parrish CC, Wangersky PJ (1987) Mar Ecol Prog Ser 35:119
149. Aaronson S, De Angelis B, Frank O, Baker H (1971) J Phycol 7:215
150. Swift DG (1980) Vitamins and phytoplankton growth. In: Morris I (ed) The physiological ecology of phytoplankton. Blackwell Scientific Publications, Oxford, p 329
151. Otterstrøm CV, Steemann-Nielsen E (1940) Rep Dan Biol Sta 44:5
152. Yariv J, Hestrin S (1961) J gen Microbiol 24:165
153. Kaartvedt S et al. (1991) Can J Fish Aquat Sci 48:2316
154. Igarashi T, Oshima Y, Murata M, Yasumoto T (1995) In: Lassus P, Arzul G, Erard E, Gentien P, Marcaillou C (eds) Harmful Marine Algal Blooms. Lavoisier, Paris, p 303
155. Igarashi T, Satake M, Yasumoto T (1996) J Am Chem Soc 118:479
156. Myklestad SM, Ramlo B, Hestmann S (1995) In: Lassus P, Erard G, Gentien P, Marcaillou C (eds) Harmful Marine Algal Blooms. Lavoisier, Paris, p 633
157. Meldahl AS et al. (1995) Ibid, p 315
158. Arzul G et al. (1993) In: Smayda TJ, Shimizu Y (eds) Toxic Phytoplankton Blooms in the Sea. Elsevier, Amsterdam, p 719
159. Honjo T (1993) Ibid p 33
160. Mingazzini M, Thake B (1995) Sci of the Total Environ 165:9
161. Alldredge AL, Passow U, Logan B (1993) Deep-Sea Res I 40:1131
162. Dundas I, Johannessen OM, Berge G, Heimdal B (1989) Oceanography, april: 9
163. Nielsen TG, Kiørboe T, Bjørnsen PK (1990) Mar Ecol Prog Ser 62:21
164. Skjoldal HR, Dundas I (1991) The *Chrysochromulina polylepis* bloom in the Skagerrak and the Kattegat in May-June 1988: environmental conditions, possible causes, and effects. Report No 175. ICES, Copenhagen
165. Graneli E, Paasche E, Maestrini SY (1993) In: Smayda TJ, Shimizu Y (eds) Toxic Phytoplankton Blooms in the Sea. Elsevier, Amsterdam, p 23

166. Lancelot C, Mathot S (1987) Mar Ecol Prog Ser 37:239
167. Lancelot C (1995) Sci of the Total Environ 165:83
168. Van Rijssel M, Hamm CE, Gieskes WWC (1997) Eur J Phycol 32:185
169. Veldhuis MJW, Admiraal W (1987) Mar Biol 95:47
170. Veldhuis MJW, Admiraal W (1985) Mer Ecol Prog Ser 26:301
171. Haug A, Myklestad S (1976) Mar Biol 34:217

CHAPTER 6

Interfacial Processes

Vera Žutić, Vesna Svetličić

Center for Marine and Environmental Research, Rudjer Bošković Institute, POB 1016, 10000 Zagreb, Croatia
E-mail: zutic@rudjer.irb.hr; E-mail: svetlicic@rudjer.irb.hr

As we have gone beyond the traditional oceanographic routines in our field studies of the characterization of sea water, we have seen that an important fraction of marine particulate matter is in the size range of only a few nanometers to hundreds of micrometers. These particles are fluid and flexible, non-living, and primarily organic. The abundance of the non-living particles greatly exceeds the abundance of living microorganisms, algae and bacteria, and viruses. Their total interfacial area in the water column surpasses by a few orders of magnitude the area of the sea surface. The structures and systems they form in sea water stand apart from the conventional colloidal system of solid particles and are described in terms of complex fluids. The mechanism of their formation, their stability, and the role they play in ocean biogeochemical cycles is becoming a most challenging exercise in converging disciplines of marine chemistry, microbiology, and biophysics.

Key words: Marine interfaces, Marine particle classes, Fluidity of marine particles, Self-assembly of biopolymers, Gel phase formation.

The Handbook of Environmental Chemistry Vol. 5 Part D
Marine Chemistry
(ed. by P. Wangersky)

1 Interfaces

Interfaces are, briefly, boundaries between different phases, the solid, liquid, and gaseous phases. The notation *interface* is the general one, whereas *surface* is used in the more restrictive meaning of a boundary with the gas phase or vacuum. "Surface" is also used when referring to the boundaries of a particle independent of what is around the particle. The general prerequisite for the stable existence of an interface between two phases is that the free energy of formation of the interface be positive [1]; were it negative or zero, the effect of accidental fluctuations would be to expand the surface region continuously and lead to eventual complete dispersion of one material into the other, as in the case of two miscible liquids. Systems involving an interface are often metastable, that is, essentially equilibrium behavior is exhibited in certain aspects although the system as a whole may be unstable in other aspects.

Interfaces derive their relevance from two related phenomena: the presence of interfacial tension and the occurrence of adsorption [2]. Qualitatively, the interfacial tension acts in any interface, trying to minimize the interfacial area. There is a free energy change (γ_l) when the surface area of a medium is increased by a unit area; the process of creating a unit area of surface is equivalent to separating two half-unit areas from contact:

$$\gamma = {}^1/_2\, W_{11}$$

where W_{11} is the work of cohesion. For solids γ_l is given in units of energy per unit area: $mJ \cdot m^{-2}$ ($erg \cdot cm^{-2}$), for liquids in units of tension per unit lengths: $mN \cdot m^{-1}$ ($dyn \cdot cm^{-1}$). If the interface is fluid, i.e., for liquid-liquid and liquid-gas boundaries, the action of the interfacial tension manifests itself in the shape that the interface assumes.

Adsorption is, briefly, the accumulation of matter at interfaces. Tendencies to adsorb vary widely between substances and interfaces. Surfactants are molecules that, because of their typical chemical composition, have a great tendency to adsorb from solution: their concentration in the interface is much higher than in the solution with which the boundary is in equilibrium. Interfacial tension and adsorption are intimately related through Gibbs' adsorption law, the most important law of interfacial science. For the simple case of one dilute uncharged component:

$$d\gamma = -\,RT\Gamma\, d\ln c$$

Here R is the gas constant, T the absolute temperature, Γ the interfacial or surface concentration in moles per unit area, and c the concentration in arbitrary units ($d\ln c = dc/c$ is the relative change in concentration and hence dimensionless).

It is more difficult to give a general definition for colloids than for interfaces, because a great variety of colloids exists and there is no consensus among workers in the field on what to call colloidal [3, 4]. Usually, the definition is made on the basis of size. A relatively general definition is as follows: a colloid is an entity having in at least one direction a dimension between 1–1000 nm. Colloids are large enough to acquire an interface (that is, the interior is different from the surrounding medium) but small enough for gravity not to be a dominant force acting upon them.

Today, as more and more information is accumulating on the properties of diverse systems at the molecular level, there is a natural desire to understand these phenomena in terms of the operative forces. In chemistry and biology, emphasis is placed entirely on the short-range force fields around atoms and molecules, rarely extending more than one or two atomic distances. Terms in molecular biology, such as molecular packing, specific binding sites, lock and key mechanisms are essentially short-range. In the different, though closely related area of colloid science, the emphasis is quite often on the long-range forces, which may determine whether two surfaces or particles are able to get close enough before they can interact via the types of short-range forces [5].

2 Interfacial Forces

In systems involving interfacial phenomena one is not so much concerned with molecular transformations as with the interactions between discrete non-bonded atoms or molecules over distances significantly greater than molecular bond dimensions. Such interactions are conventionally called physical interactions, resulting from physical forces and producing physical bonds. While physical interactions may perturb the electronic configurations of the molecules or atoms involved, the electrons themselves remain associated with their original systems. Physical bonds usually lack the specificity, stoichiometry, and strong directionality of covalent bonds. They are therefore the ideal candidates for holding molecules together in liquids, since the molecules can move about and rotate while still remaining "bonded" to each other. Nevertheless, physical binding forces can be as strong as covalent bonds, and even the weakest is strong enough to hold all but the smallest atoms and molecules together in colloidal and biological assemblies. These properties, coupled with the long-range nature of physical forces, make them the regulating forces in all phenomena that do not involve chemical reactions. Many important systems and processes, such as self-assembly in the aquatic environment, form as a result of physical, inter- and intramolecular interactions. Such assemblies and conformations exist because the physical forces binding them together operate over distances greater than those of covalent bonds yet hold various molecules at a proper distance and with the proper strengths so that they can successfully carry out their vital functions.

Four main type of forces act between surfaces in liquids: van der Waals, electrostatic, solvation (hydration), and steric forces. For a typical colloidal system of rigid particles in water, it is rare for more than two of these forces to be dominating the interaction at any one time. In contrast to this, the forces between highly mobile amphyphilic surfaces of fluid bilayers and biological membranes can have all four operating simultaneously, as well as other – more specific – types of interaction. Hydrophobic force can be far stronger than the van der

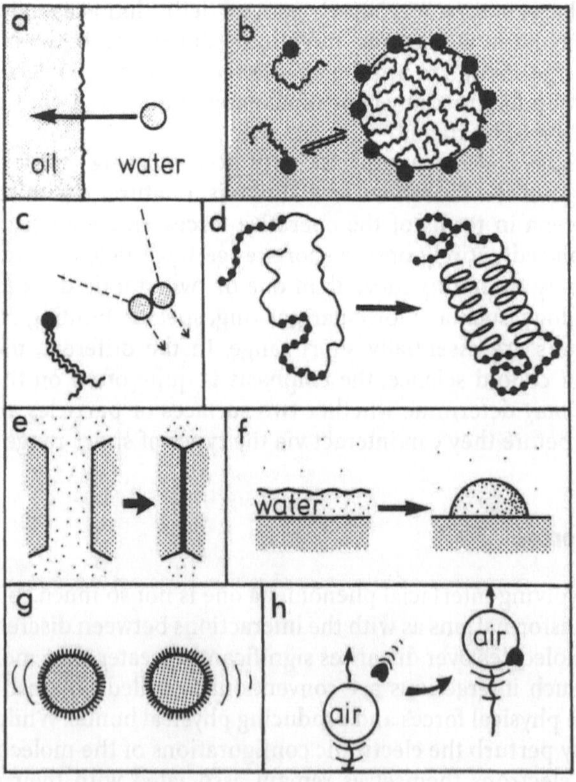

Fig. 1a, b. Examples of attractive hydrophobic interactions in aqueous solutions. **a** Low solubility/immiscibility; **b** micellization; **c** dimerization and association of hydrocarbon chains; **d** protein folding; **e** strong adhesion; **f** non-wetting of water on hydrophobic surfaces; **g** rapid coagulation of hydrophobic or surfactant coated surfaces; **h** hydrophobic particle attachment to rising air bubbles (reproduced from Israelachvili [5])

Waals attraction, especially between hydrocarbon surfaces. At separation below 10 nm the hydrophobic force appears to be insensitive to changes in the type and concentration of electrolyte ions in the solution. The absence of the screening effect by ions attests to the non-electrostatic origin of this interaction. In solutions containing divalent cations such as Ca^{2+}, it can continue to exceed the van der Waals attraction out to a separation of 80 nm. The long-range nature of hydrophobic interaction has a number of important consequences (Fig. 1). It accounts for the rapid coagulation of hydrophobic particles in water and may account for rapid folding of proteins.

3
Neglected Dimension of Marine Interfaces

As has recently emerged from field studies using sea water characterization beyond the traditional oceanographic routines [6–9], an important fraction of marine particles is in the size range of only a few nanometers to hundreds of micrometers. The particles are fluid and flexible. Their total interfacial area in the water column surpasses by a few orders of magnitude the area of the sea surface [9]. Interfacial structure, function, and chemistry of marine particles meet at a resolution near 0.001 μm.

Fluid or flexible particles could be grouped with small molecular aggregates such as micelles, bilayers, vesicles, and biological membranes that form readily in aqueous solution by spontaneous self-association or self-assembly [10] of certain amphiphilic molecules. These structures and systems they form – sometimes collectively referred to as association colloids or complex fluids [11] – stand apart from the conventional colloidal particles. This is because the forces that hold amphiphilic molecules together in micelles and bilayers are not due to strong covalent or ionic bonds but arise from weaker van der Waals, hydrophobic, hydrogen-bonding, and screened electrostatic interactions. Thus, if the solution conditions, such as the electrolyte concentration or the pH, of an aqueous suspension of micelles or vesicles are changed, not only will this affect the interactions between the aggregates but it will also affect the intermolecular forces within each aggregate, thereby modifying the size and shape of the structures themselves.

The new classes of highly abundant particles, which had remained undetected by traditional oceanographic techniques [12] because of their small size, consistency, or transparency, were recently identified using electronic and electrochemical particle counters, electron microscopy, and histological staining techniques. These new particle classes are defined operationally in Table 1.

They were described as fluid, flexible, fragile, deformable and filmy. The abundance of non-living particles greatly exceeds the abundance of living microorganisms, algae and bacteria, and viruses. We shall describe the discoveries of fluid particle populations in sea water, all having properties of complex fluids, in order as they appeared in literature.

3.1 Fluid Surface-Active Aggregates and Adhesion at Model Interface

Electrochemical measurements of surfactant activity in fresh unfiltered samples of estuarine [16, 17] and sea water [18] revealed the presence of surface-active particles. This highly reactive and hydrophobic fraction of particles was not amenable to analysis by conventional methods. Using a dropping mercury electrode as a sensor directly immersed in fresh unfiltered samples, Žutić et al. detected a random appearance of coalescence signals of polydispersed surface-active constituents, resembling oil droplets. Adhesion and subsequent spreading of a fluid particle causes displacement of double layer charge at the mercury/sea water interface (Fig. 2), which can be recorded as a well resolved electrical signal on the millisecond time scale [23]. The adhesion force can be fine-tuned [24] by changing the electrode potential with all other properties of the system remaining the same (Fig. 3). In such a way the interplay between electrostatic and hydrophobic interactions can be sorted out.

At the positively charged interface the adhesion was most favorable and the duration of coalescence signals was between 10 and 100 ms. The signal ampli-

Table 1. Major classes of non-living particles in the upper ocean (euphotic layer)*

Type	Size	Location	Detection method	Concentration N/L	Composition	Description
Small colloids [7, 13, 14]	5–200 nm	Santa Monica Basin, North Atlantic, Northwest Pacific	TEM	10^7–10^{12} maximum at thermocline	largely organic rounded,	rounded, globular, shape, diffuse
Submicrometer particles [6, 15]	0.38–1 μm	North Pacific off Japan, Northwest Atlantic Schlep	Coulter particle counter	5×10^{10}–8×10^{10} in top 40 m, $< 10^9$ below 200 m	organic with high water content	highly, flexible, mostly, amorphous
Fluid surface-active aggregates [16–18]	1–100 μm	Mediterranean	adhesion at mercury electrode	10^5–5×10^7 maximum at halocline	organic, surface-active,	fluid, hydrophobic
Transparent exopolymeric particles (TEP) [8, 19]	3–100 μm	off California coast	microscopy after Alcian blue staining	3×10^4–5×10^6	polysaccharide matrix with high water content	flexible, deformable, “filmy”, gel-like
Protein containing particles (Coomassi stained), CSP [9]	2–500 μm	off California coast	microscopy after Coomassie blue staining	10^6–10^8	organic proteinaceous	flexible, deformable gelatinous
Giant aggregates (mucilage) [20–22]	> 1 m	North Adriatic, episodic	scuba-diving	0–10 in 100 m^3 above thermocline (~20 m)	Organic (1‰), polysaccharides	Mucus aggregates, gel phase

* Defined operationally according to measurement techniques.

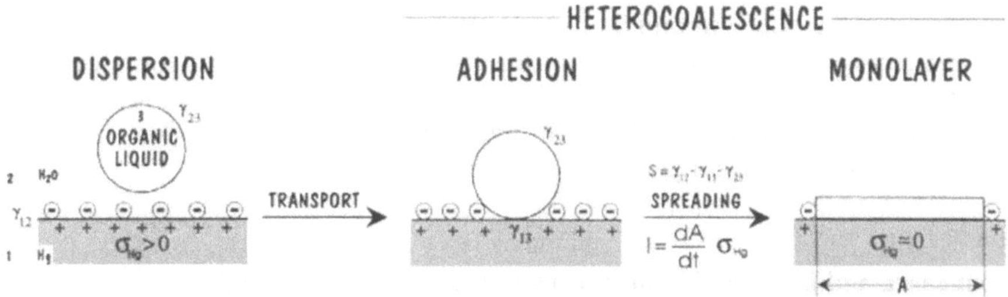

Fig. 2. Schematic diagram of the attractive interaction between dispersed organic particles and a positively charged mercury electrode in an aqueous electrolyte solution (where σ_{Hg} is surface charge density, S spreading coefficient, I electrical current, dA/dt rate of spreading into a monolayer film; reproduced from Žutić et al. [23])

Fig. 3 a–d. Control of adhesion force by electrode potential. Photographs of an *n*-hexadecane droplet, deposited at mercury electrode/aqueous electrolyte interface at −400 mV, taken at constant potentials of **a** −550, **b** −1300, **c** −1400, and **d** −1450 mV. Magnification 10× (reproduced from Ivošević and Žutić [24])

tudes corresponded to equivalent spherical diameters of 1 to 100 µm. The lower detection limit was given by the instrumental noise, and by the fluidity of particles – solid particles were not detectable.

The highest abundance of fluid particles, up to 10^8 L^{-1}, was found in estuarine mixing zones and at the sea surface [17, 18]. The fluid particles were described as surface-active aggregates formed by condensation of excretion and decomposition products of phytoplankton – polysaccharide, proteinaceus, and lipid components, assembled in a form of an insoluble surface-active complex. The condensation process is suggested to be salting out and supramolecular organization by hydrogen-bonding and hydrophobic attraction under a salinity gradient and shear stress [25].

3.2
Abundance and Reactivity of Submicrometer (0.38–1.0 μm) Particles

Koike and coworkers in 1990 [6] reported the discovery of a hitherto unsuspectedly abundant population (10^{10}/L) of non-living particles of submicron dimension with "unusual properties". Examining oceanic particles with a spherical diameter between 0.38–1.0 μm using a resistive-pulse particle counter, they identified highly flexible non-living particles, together with the otherwise standard populations of viruses, bacteria, cyanobacteria, and small eukaryotes. More than 95% of the submicrometer particles were non-living. These particles are very fragile – ultrasonication reduces drastically the number of particles – and flexible enough to sneak through pore sizes nominally only one-quarter of their diameter.

The new particles were characterized as organic with a high content of water, as inferred from a very small difference in density between the particles and the surrounding sea water. Particles readily aggregated by agitation or surface coagulation [26] in a bubbling column. Upon ultracentrifugation above 100,000 g, they progressively coalesced into a film composed of a diffuse, electron-transparent material. The measurement techniques did not, however, provide an estimate of the carbon content or mass of material in the submicrometer pool.

The submicrometer particles appeared to differ significantly from particles previously thought to be the smallest entities formed by microbial processes such as mortality. Nevertheless, the authors viewed their formation mechanism as a microbial process and not a physico-chemical aggregation of smaller entities. A short residence time of organic matter pool in the same size range (10 days in the upper ocean) was measured by radiotracer techniques [27, 28]. Hence, a rapid aggregation of submicrometer particles into larger size fractions may be inferred. This particle population provides a large interface area, of the order of 20–50 m^2 per cubic meter of sea water.

3.3
Occurrence and Aggregation of Small Colloid Particles (Nanoparticles) in Sea Water

Theoretical arguments combined with data from laboratory experiments have suggested that the aggregation of marine colloidal matter is an important mechanism for transferring dissolved substances into macroparticles, but there had been little evidence that colloid aggregation is significant in the ocean. The major advancement in characterizing the colloidal state in the ocean came from a further step down towards measuring particles at a smaller size scale. In 1991 Wells and Goldberg [7] reported the discovery of a large population of particles smaller than 120 nm in size. The colloids were collected by ultracentrifugation

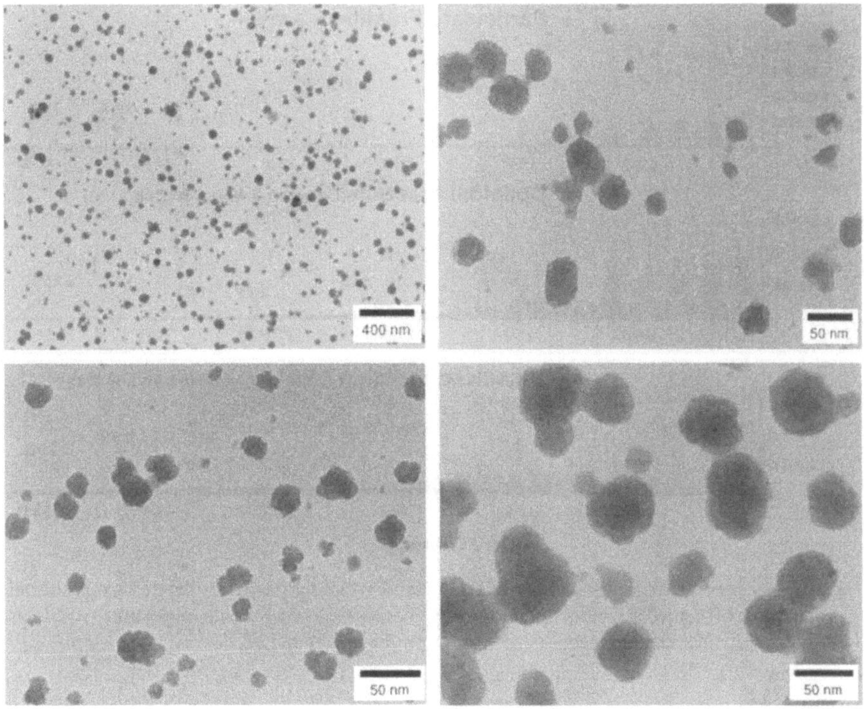

a c b d

Fig. 4a–d. First transmission electron micrographs of marine colloids. Sea water samples were collected at a station in Santa Monica basin, September and July 1990 at different depths: **a**, **b** 75 m; **c** 400 m, and **d** 850 m (reproduced from Wells and Goldberg, 1991 [7])

directly on specimen grids for transmission electron microscopy that was combined with energy-dispersive X-ray spectroscopy (Fig. 4). The small colloidal particles were at least three orders of magnitude more abundant than larger submicron particles. The abundance of small colloidal particles reaches 10^{12} particles per liter with a total surface area in the order of 10 m^2 per cubic meter.

The marine colloidal matter was characterized primarily as organic. Its organic components differ but slightly from its dissolved and particulate counterparts (Fig. 5) and in principle could all be derived from the same general source [29]. The marine colloids contained primarily carbohydrates, fatty acids, minor amounts of proteinaceous compounds, and electropositive elements – iron and aluminum. The colloids have a granular structure – the granule size ($\approx$2 nm) [30] suggests that the colloids are aggregates composed of relatively low molecular weight organics. One of the primary sources of marine colloids is likely the agglomeration of some fraction of the truly dissolved organic phase.

The fractal dimensions of aggregates characteristically indicate highly variable colloidal dynamics in sea water [13]. Colloid numbers increase nearly logarithmically with decreasing size, creating a continuum of particle sizes that links nanometer-size matter to large particles, which ultimately remove matter from the ocean by sinking.

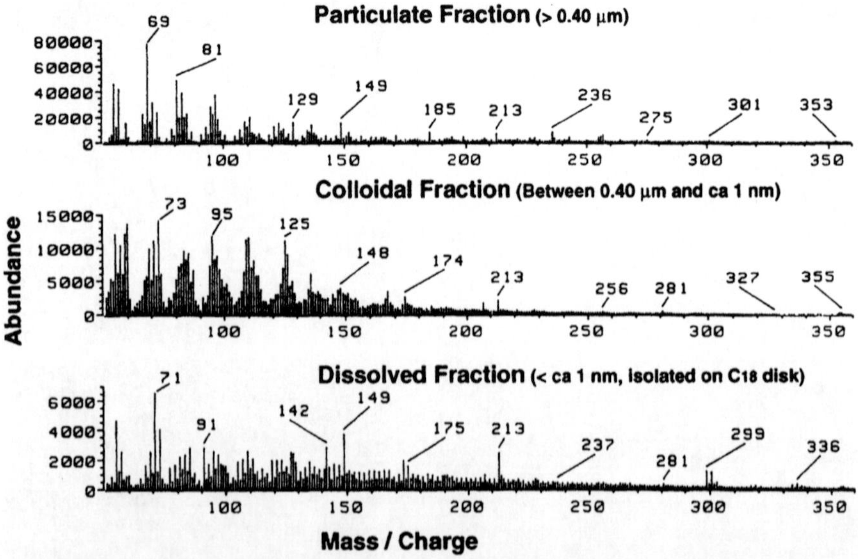

Fig. 5. Mass spectra from dissolved, particulate and colloidal particles isolated from coastal surface water off La Jolla, California, June 1993. Identical 50–360 atomic mass units windows were chosen for the detail (reproduced from Vernonclark et al. [29]

The removal and residence time of colloidal substances from sea water would depend upon either their degradation to soluble components or their aggregation to form sinking macroparticles. The results of Wells and Goldberg support the concept of a continuous flow of material between the colloidal, particulate, and dissolved states by aggregation. The microbial loop may be less important as a direct sink for colloidal organic carbon.

3.4
Transparent Exopolymeric Particles

Polysaccharide-specific staining revealed the existence and high abundance of a class of large, discrete, transparent particles [8, 19] that were a missing component in fitting the theoretical models to the aggregation of algal blooms in nature [31–34]. Transparent exopolymer particles (TEP) ranged from 28 to 5000 particles per mL sea water and ranged from 3 to >100 µm in the longest dimension at coastal stations off California. In 1965, Wangersky described aggregates newly formed from filtered sea water as "looking like bits of cellophane" [35]. While the existence of transparent marine particles had been reported previously [36], these descriptive accounts have not received due attention because the abundance and significance of the particles had remained unknown. Particles indiscernible by light microscopy or the electronic particle counter, but made visible by Alcian blue staining [37], were present in all natural sea water samples. The strong reaction of particles with Alcian blue indicated that these particles contained abundant polysaccharides. Most stainable particles were discrete, highly deformable films, discs or strings, containing no or a few visible inclusions. TEP were also highly significant components of larger aggregates. The disruption of aggregates by EDTA indicated the involvement of bivalent cation bridges.

TEP originate from dissolved polysaccharide exudates released by phytoplankton and bacteria. The existence of microbial exudates acting as large, discrete particles, rather than as dissolved molecules, or as coating on other particles, suggests that the transformation of dissolved organic matter into particulate form in the sea can occur via a rapid abiotic pathway. TEP appear truly particulate in that they are retained on filters, harbor attached bacteria, and are subject to aggregation into larger particles via physico-chemical aggregation. On filters TEP appeared as two-dimensional sheets and films of discrete particles, while in three dimensions TEP are likely to be highly fractal aggregations of smaller microfibrils and colloids with a high interstitial water content, resembling a gel.

4 Self-Assembly of Marine Organic Matter Into Polymer Gels

The concept of sea water as a gel-like polymeric matrix with colloids and particles embedded as suprapolymeric "hotspots" comes from the experience of marine microbial ecology [38] (Fig. 6).

Gels [39, 40] are a spectacular form of soft matter – responding to very weak external perturbations. They swell or contract under minute changes of certain control parameters: adding salt, changing pH, and changing temperature or solvent. The gel is a state of matter, neither liquid nor solid, or conversely both liquid and solid. As a result it possesses both the cohesive properties of solids and the diffusive properties of liquids. Any gel consists of at least two components: a molecular network, which is swollen by a liquid. Network structures of organic polymeric gels can be obtained as a result of chemical or physical processes. Covalent cross-linking leads to the formation of irreversible gels. Cross-links resulting from weaker interactions (hydrogen bond, ion complexation, van der Waals' forces) can be broken, and their number and strength depend on thermodynamic and mechanical parameters; accordingly gels are reversible. When formed by physical processes, the tie-points often result from a complex organization of macromolecules (e.g., formation of mesophases and small colloids).

A recent study of Chin et al. [41] demonstrated that polymers in sea water indeed form a gel. In their laboratory experiment they found that *DOM* polymers in 0.22 μm-filtered sea water spontaneously assemble to form polymer gels ranging from colloidal to micrometer size. Fresh filtrates contained a polydisperse set of polymers with sizes ranging from 2 to 200 nm. The assembly process is not linear and the increase of particle size followed second order kinetics. The equilibrium size of 5 μm (Fig. 7) was reached in 50–80 h. The equilibrium particle size of the assembled gels depended on the total amount of available free polymers in the 10 mL sample. In the ocean, with virtually infinite source of free polymers, the gels could potentially reach larger sizes.

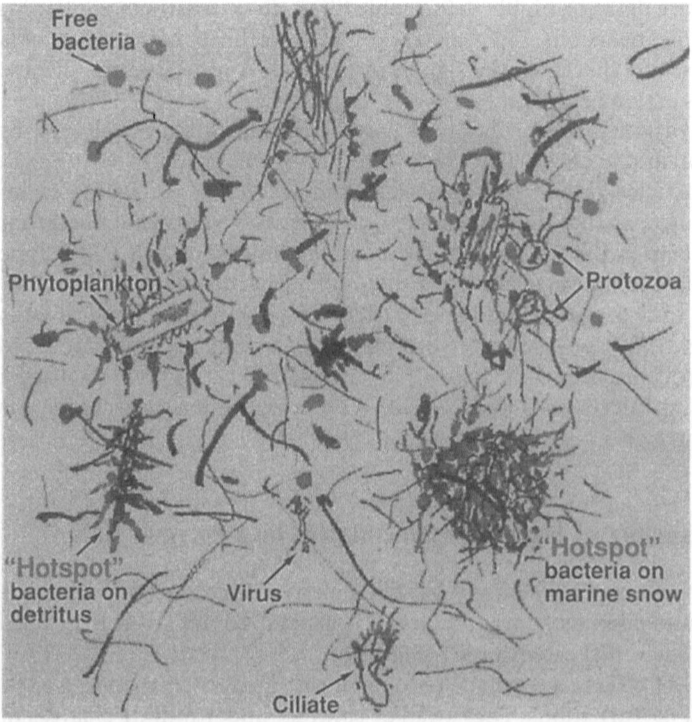

Fig. 6. Sketch of sea water as an organic matter continuum: a gel of tangled polymers with embedded strings, sheets and bundles of fibrils and particles, including living organisms, as "hot spots" (reproduced from Azam [38])

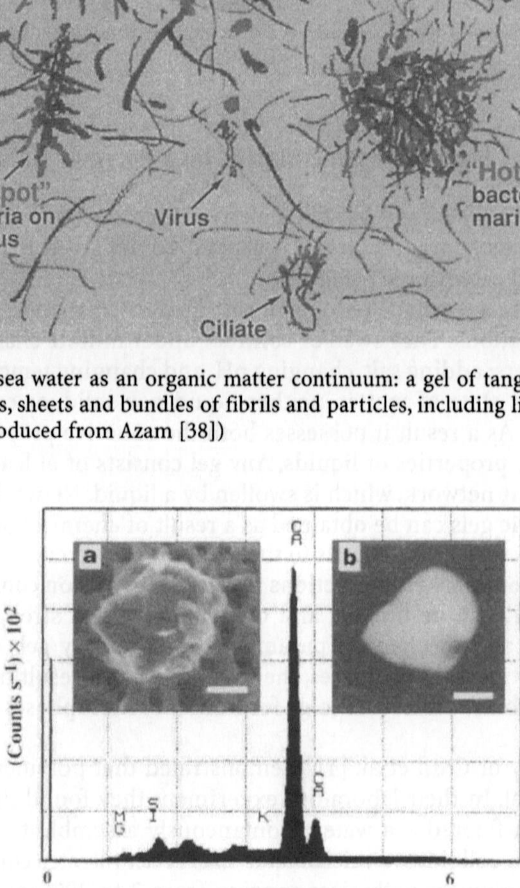

Fig. 7. Averaged X-ray spectra showing the elemental composition inside a marine microgel with the large peak of calcium and the absence of phosphorus. **Inset a** shows an environmental SEM micrograph of a hydrated marine microgel (scale bar, 2 μm). **Inset b** reveals the presence of bound calcium in a microgel stained with chlortetracycline (reproduced from Chin et al. [41])

Ion bridging between biopolymers by Ca^{2+} and Mg^{2+} ions seems to be the primary driving mechanism. The aggregates are composed of carbohydrates, proteins, and lipids, similar to the colloidal concentrates of Wells and Goldberg [29]. This suggests that the stability of organic substances in sea water may largely depend on their interfacial characteristics and to a lesser extent on their chemical composition. Because the probability of polymer gels coalescing and becoming stable increases in proportion to the square of polymer lengths, the smallest organic constituents in sea water would be less likely to be incorporated into macropolymer assemblies than would colloidal constituents, as revealed from the recent data on the age of colloidal and soluble dissolved organic carbon in the upper ocean [27, 28, 42].

The even more surprising result of the laboratory experiment is that particulate gels re-form from soluble and colloidal precursors after previous gels are removed, as a consequence of re-equilibration in the system [41]. The phenomenon itself is strikingly similar to already classical experiments with organic aggregate formation in filtered sea water [35, 43]. The biogeochemical implication would be that colloid aggregation in the ocean might be strongly influenced by particulate removal rates.

The missing component in these experiments is the absence of microbial activity, so it is still not possible to assess the relative importance of competing processes of biodegradation and spontaneous assembly into gel phase [44] (Fig. 8).

Phytoplankton and bacterial production are not always tightly coupled so that the input of colloids may outpace degradation over shorter time intervals.

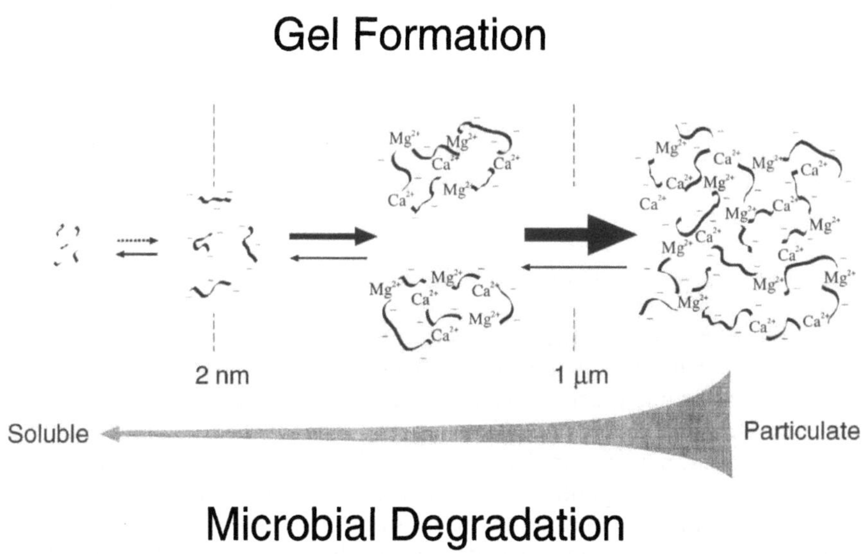

Fig. 8. Representations of the dynamic interplay between marine gel formation and microbial degradation in sea water. The negative charges depict the anionic nature of marine organic matter while cationic bridging appears to drive polymer coalescence. Thickness of arrows indicates relative reaction rates (reproduced from Wells [43]

Fig. 9. Northern Adriatic giant gel aggregate at 10 m depth as captured by a scuba-diver in August 1997 (courtesy of Gerald Müler-Niklas)

Such a situation may be exemplified by the Northern Adriatic phenomenon of massive giant aggregate formation (Fig. 9) that was preceded by a high density of fluid surface-active aggregates in the whole water column above the thermocline [22]. On the basis of long-term electrochemical measurements of surfactant activity in sea water samples in different seasons [18, 45, 46] of the abundance of fluid surface-active aggregates and the characterization of the gel phase of giant aggregates during the summer of the 1991 and 1997 episodes, we put forward a plausible scenario with the sequence of steps leading to massive macroaggregate formation and its evolution when the physico-chemical interactions and the bacterial processes are placed together [22].

4.1 Massive Macroaggregation in the Northern Adriatic

Eutrophication of the Northern Adriatic, due to run-off from the Po and other rivers, causes hyperproduction of microalgae during spring/summer at rates which far exceed the grazing potential of herbivores [47]. Consequently, large standing stocks of microalgae build up and extracellular polymers, mainly polysaccharides, may accumulate as a colloidal suspension (colloidal organic matter, COM) in the euphotic layer above the picnocline of the highly stratified water column. Bacteria may not themselves be the major producers of extracellular polymers, but through their control of phytoplankton aggregation they may influence the size of phytoplankton blooms and the sinking carbon flux.

With prolonged residence time the accumulated colloidal organic matter undergoes aggregation over the whole water column above the thermocline to

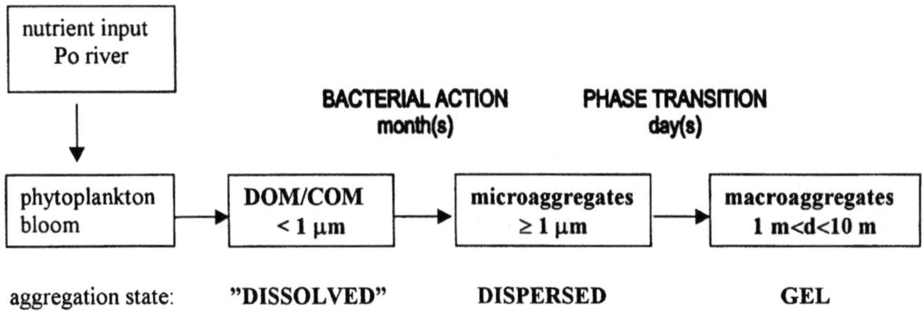

Scheme 1. Schematic presentation of the main steps leading to macroaggregate formation

microagreggates of the size range 1–100 μm. When the microaggregates attain the critical concentration N_c (for which our present estimate centers around $N_c \approx 5 \times 10^7$ L^{-1} and the critical interfacial properties, the large-scale phase transition from the dispersed to the gel state takes place Scheme 1).

The gel phase contains less than 1 ppm of organic carbon [48, 49] composed primarily of polysaccharide matrices. The gel phase is also a site of intensive bacterial activity. The adhesion behavior of gelatinous macroaggregate material revealed the coexistence of hydrophilic and hydrophobic domains. Surfactant activity in the gel was up to 50 times higher than in the adjacent sea water.

The sol-gel transition is almost instantaneous. The shape and the size of macroaggregates formed by the phase transition are dictated by the local hydrodynamics (turbulence and shear) and the proximity of sea surface and density interfaces in the water column. Buoyancy and vertical migration of macroaggregates depend upon the ongoing microbial activity within the gel. The persistence and transient stability of macroaggregates is governed by: (1) collapse of the gel matrix due to a gel-solid transition, which is a slow process (and subsequent sinking below the picnocline) and (2) large-scale physical processes, such as strong winds and currents, that cause physical disaggregation and export from the aquatorium.

5
Concluding Remarks

We have presented a new aspect of marine interfacial processes emerging from recent reports on existence of fluid and flexible particles in sea water, in the size range of 2 nanometers to several hundreds of micrometers, and even meters, and having the largest interfacial area in the upper ocean.

Fluid state could be assigned to all size classes of these particles and also a primarily organic composition – polysaccharides, proteins, and lipids, with a high water content. It appears that the particle formation mechanism and creation of new interfaces is based on the self-organization of biopolymers into complex fluid structures, rather than on adsorption onto preexisting inorganic interfaces. The primary driving force in the supramolecular organization was suggested to be hydrophobic interaction and ion bridging with Ca^{2+} and Mg^{2+}.

The reported fast formation rate of microscopic gel particles from biopolymers under abiotic conditions and the reversibility of this reaction in seawater has far reaching implications.

Being able to identify more clearly the characteristics of abiotic birth and growth of new interfaces in sea water by self-assembly, we can take one step further and try to assess their importance relative to biological production and biodegradation of particles in the ocean.

6
References

1. Adamson AW, Gast AP (1997) Physical chemistry of surfaces, 6th edn. Wiley-Interscience, New York
2. Lyklema J (1991) Fundamentals of interface and colloid science, Academic Press, London
3. Stumm W, Morgan JJ (1996) Aquatic chemistry, 3rd edn. Wiley-Interscience New York
4. Buffle J, van Leeuwen HP (eds) (1992) Environmental particles, vol 1, IUPAC Environmental analytical chemistry series. Lewis, Boca Raton
5. Israelachvili J (1992) Intermolecular and surface forces, 2nd edn. Academic Press, London
6. Koike I, Hara S, Terauchi K, Kogure K (1990) Nature 345:242
7. Wells ML, Goldberg ED (1991) Nature 353:342
8. Alldredge AL, Passow U, Logan BE (1993) Deep-Sea Res I 40:1131
9. Long RA, Azam F (1996) Aquat Microb Ecol 10:213
10. Thermodynamic principles of self-assembly and the corresponding thermodynamic equations can be found in chaps 16, 17 of [5]
11. Smalley MV (1996) Long range attractions in charged colloids. In: Arora AK, Tata BVR (eds) Ordering and phase transitions in charged colloids, Complex fluids and fluid microstructure series. VCH, New York, chap 12
12. Oceanographers traditionally employ filters 0.2–07 μm to arbitrarily partition sea water into "dissolved" and "particulate" phases in order to assess the amounts of constituents associated with sinkable particles
13. Wells ML, Goldberg ED (1993) Mar Chem 41:353
14. Wells ML, Goldberg ED (1994) Limnol Oceanogr 39:286
15. Longhurst AR, Koike I, Li WKW, Rodriguez J, Dickie P, Kepkay P, Partensky F, Bautista B, Ruiz J, Wells ML, Bird DF (1992) Deep-Sea Res 39:1
16. Žutić V, Pleše T, Tomaić J, Legović T (1984) Mol Cryst Liq Cryst 113:131
17. Žutić V, Legović T (1987) Nature 323:612
18. Marty J-C, Žutić V, Precali R, Saliot A, Cosović B, Smodlaka N, Cauwet G (1988) Mar Chem 26:313
19. Logan BE, Passow U, Alldregde AL, Grossart HP, Simon M (1995) Deep-Sea Res II 42:203
20. Stachowitch, M, Fanuko N, Richter M (1990) Mar Ecol 5:243
21. Leppard GG (1995) In: Vollenweider RA, Rinaldi A (eds) Marine mucilages. Sci Total Environ 165:1
22. Ivošević N, Svetličić V, Kovač S, Kraus R, Žutić V, Furić K (1998) Ocean Science Meeting, San Diego USA
23. Žutić V, Kovač S, Tomaić J, Svetličić V (1993) J Electroanal Chem 349:173
24. Ivošević N, Žutić V (1998) Langmuir 14:231
25. Žutić V, Tomaić J (1988) Mar Chem 23:51
26. Johnson BD, Zou X, Wangersky PJ (1986) Neth J Sea Res 20:201
27. Moran SB, Buesseler KO (1992) 359:221
28. Santchi PH, Guo LD, Baskaran M, Trumbore S, Southon J, Bianchi TS, Honeyman B, Cifuentes L (1995) Geochim Cosmochim Acta 59:625
29. Vernonclark RN, Goldberg ED, Bertine KK (1995) Chem Ecol 11:69
30. Wells ML, Goldberg ED (1992) Mar Chem 40:5

31. Alldredge AL, McGillivary P (1991) Deep-Sea Res 38:431
32. Hill PS (1992) J Geophys Res 97:2295
33. Johnson BD, Kranck K, Muschenheim DK (1994) Physicochemical factors in particle aggregation. In: Wotton RS (ed) The biology of particles in aquatic systems, 2nd edn. Lewis Publishers, Boca Raton, chap 4
34. Alldredge AL, Jackson GA (eds) (1995) Aggregation in marine systems. Deep-Sea Res II 42:1–273
35. Wangersky PJ (1965) Amer Sci 53:358
36. Gordon DC (1970) Deep-Sea Res 17:175
37. Parker BC, Diboll AG (1966) Phycologia 6:175
38. Azam F, (1998) Science 280:694
39. Vold RD, Vold MJ (1983) Colloid and interface chemistry. Addison-Wesly, London, chap 16
40. Cohen Addad JP (ed) (1996) Physical properties of polymeric gels, Wiley, Chichester
41. Chin W-C, Orellana MV, Verdugo P (1998) Nature 391:568
42. McCarthy M, Hedges J, Benner R (1996) Mar Chem 55:281
43. Riley GA (1970) Adv Mar Biol 8:1
44. Wells ML (1998) Nature 391:530
45. Žutić V, Ćosović B, Marčenko E, Bihari N, Kršinić F (1981) Mar Chem 10:505
46. Ćosović B, Žutić V, Vojvodić V, Pleše T (1985) Mar Chem 17:127
47. Vollenweider RA, Marchetti R, Vivani R (eds) (1992) Marine coastal eutrophication. Sci Total Environ [Suppl]
48. Ćosović B (1997) personal communication
49. Müller-Niklas G, Schuster S, Kaltenböck E, Herndl GJ (1994) Limnol Oceanogr 39:58

CHAPTER 7

Intercomparisons and Intercalibrations

Peter J. Wangersky

School of Earth and Ocean Sciences, University of Victoria, P.O. Box 3055, Victoria, B.C., V8W 3P6, Canada
E-mail: wangers@ibm.net

The institution of large-scale international cooperative research programs in the study of ocean systems has made more rigorous quality control over analytical procedures necessary. A series of intercalibration and intercomparison exercises have shown that increased care is needed in running even those determinations which have been run for years, and are considered as "standard" methods. Care in sampling is needed to ensure that the various samplers used are sampling for the same materials. At this time, we do not know the scale of heterogeneity to be expected in the open ocean, and so must eliminate as many sources of error as possible from our analytical and sampling schemes, since what differences are left will be assigned to the environment.

Keywords: Intercalibration, Intercomparison, Sampling, Heterogeneity.

The Handbook of Environmental Chemistry Vol. 5 Part D
Marine Chemistry
(ed. by P. Wangersky)

1
Introduction

The major task of the marine chemist has been to discover the mechanisms controlling the cycling of the elements from land through the oceans and into the deep sea sediments. In the early days of the science, the largest component of the effort was involved simply in devising analytical methods for the elements of choice; those elements in reasonable concentration, such as chlorine, present as chloride ion, had to be determined to accuracies and precisions greater than normally required for any routine procedure in order to fulfill the needs of the physical oceanographers, while the elements present in lesser quantities were at increasingly small concentrations, and in a matrix full of interfering substances. The possible sources of contamination in the sampling and sample-handling process were far from understood.

When I first encountered the field, in the late 1940s, marine chemists or chemical oceanographers were thin on the ground, and had largely been recruited from the ranks of analytical chemists. The chemistry run routinely on most cruises was that required by the physical and biological oceanographers: salinity, dissolved oxygen, phosphate, and occasionally nitrate and alkalinity, a suite of tests which would have been familiar to the scientists on the *Challenger* expedition. They would have felt quite at home with most of the equipment used, as well; only the filter photometer or the spectrophotometer would have been new and strange. Even the magnetic stirrer was still some years in the future. Some work was being carried out on trace metals, usually as part of a Ph.D. thesis, but, in the absence of available positions in the field, this work was seldom carried much beyond the thesis level. A little work had been done on the dissolved organic carbon (DOC) content of seawater, but the amounts found had been so small that the DOC content was generally considered to play no important role in the geochemical cycles of the oceans.

With the numbers of marine chemists so small, it was indeed unlikely that any two of them would be working on the same elements at the same time; thus there was not only little need, but also little possibility, for intercomparison or intercalibration. In addition, only a handful of institutes were studying the open ocean, so that it was extremely unlikely that expeditions would overlap in time and space. In the United States, it was generally conceded that the Pacific Ocean belonged to Scripps and the Atlantic to Woods Hole. When significant differences between oceanic regions appeared, it was assumed these were real differences.

All of this changed with the advent of the International Geophysical Year (IGY) in 1956. This was a multinational project with a large oceanographic component, involving ships of many nations and carefully planned, overlapping cruise patterns, designed to cover the world oceans. Theories of ocean circulation were propounded based not only on the differences between different areas of the oceans, but also between differences between analyses made during IGY and those made in the same areas many decades earlier. It was obvious to anyone of an untrusting nature that in order for the data from such large, multinational, multiship projects to be useful, some higher degree of quality control had to be imposed on the data collections.

Since that time, huge international projects – GEOSECS, JGOFS, STIE, PEX'86 - have become a regular feature of oceanographic research, a way to find the funding necessary to support ship time, instrumentation, and extra research personnel not otherwise available to the individual researcher, as well as a way to achieve the degree of near-simultaneous coverage necessary to elucidate the large scale, not necessarily constant patterns of oceanic circulation. In order to make any sense at all of the accumulating data, the accuracy, precision, and comparability of the work done by many scientists and technicians, from a collection of countries differing in levels of technical training and available instrumentation, and using methods of varying levels of sophistication, must be evaluated. It is in this context that intercomparability and intercalibration exercises have become a normal part of the routine of most large oceanographic departments and institutes.

I would propose that before one of these large databases accepts input from a new member, three questions need to be answered. The first, and possibly the simplest, is when given a bottled sample to be analyzed, how well does the laboratory measure what is in the bottle? This is the substance of the normal intercalibration exercise, and should be part of the quality assurance/quality control procedure of any analytical laboratory, industrial or research, running large numbers of standard analyses.

The second question is does this method measure what we think it is measuring? When good laboratories produce discrepancies in intercalibration exercises, the tendency in each laboratory is to consider the others to have had an off-day, at best (and to be staffed by incompetents, at worst). If two laboratories are running different methods, or, in these days of black-box analysis, if they are using different machines, careful comparison of results on a number of samples of differing composition may reveal that the methods are in fact not measuring the same quantities. Often this kind of intercomparison is best run by tooling up to use both methods in the same laboratory, or by taking several "black-boxes" on the same oceanographic cruise.

The third question, the one least well answered, is are the sampling methods employed by the several institutes involved actually sampling the same universe? We tend to forget that the process starts not with the bottled sample, but with the parcel of water still floating freely in the ocean. Every action we take in isolating, lifting, and bottling the water can change its composition. The first law of oceanographic research is still "As soon as you have decided on a method of sampling, you have begun to bias your data". What work has been done on comparison between sampling methods has largely been done on a single ship or laboratory basis; more relevant to the multinational projects are the few between-ship comparisons. In the following sections, we will examine more closely each of these questions in turn.

2
Intercalibrations – Can We Measure Them?

When we begin to calibrate an instrument or an analytical method, we generally start with a simplified version of the medium we will be analyzing, with samples covering the whole expected range of the substance to be measured. In the early stages of an investigation into trace quantities in the marine environment, this step can in itself lead us astray, since we may not be altogether certain of the true values of the materials involved. The possibilities for contamination in the sampling process were not fully appreciated in the early trace-metal analyses in seawater; as a result, some of the values given in the early literature are somewhat higher than the currently accepted values. For example, before we understood the importance of using nonmetallic or plastic-coated sampling cables, our values for total and particulate iron in seawater were undoubtedly too high and random [1].

Once we are certain that our method is supplying reasonable answers with our simplified medium, we can test it on seawater samples with careful additions of the substance in question. The retrieval of the proper numbers from this procedure can furnish us with a good deal of confidence in the relative accuracy of our method, but we would still like to have a fixed point of reference, an analysis of a sample whose concentration of the substance in question has been determined by a "referee" method. This need can be filled by the use of standard reference materials. Excellent discussions of the use of these materials as applied specifically to marine samples can be found in the literature [2–4].

For most industrial uses, the relevant materials can be secured from some national or international source; in the United States, for example, a wide variety of reference materials can be secured from the National Institute of Standards and Technology (<url>http://ts/nist.gov/</url>). Until fairly recently, however, little was available in the way of marine materials from this source and, even now, the available materials are most plentiful in the area of organic pollutants. Some of the most useful reference materials, particularly for trace-metal analysis in seawater, have been furnished by the Institute for National Measurement Standards of the National Research Council of Canada (<url>http://www.ems.nrc.ca</url>). Biological and geological materials are also available from this, and from several other sources. The use of such reference materials should enable analysts to evaluate the performance of their laboratory and of their methods before chancing their hand at the public exposure of the intercomparison round-robin. However, if the laboratory expects to run any considerable number of samples as a regular routine, neither the occasional intercalibration round-robin nor the infrequent running of reference materials can replace a well-considered quality assurance/quality control program.

Unfortunately, such programs have always been the exception rather than the rule in academic laboratories, at least in part because there is no obvious source of funding for them. Academic laboratories also suffer from frequent changes of personnel, as graduate students obtain their degrees and seek more permanent employment; it is the rare academic who has a steady enough source of funding to retain experienced technicians.

Furthermore, an academic laboratory is not likely to set up a program which involves long-term monitoring of a particular area for a given suite of elements or compounds. There have been exceptions to this rule but, in general, the research aims of academic laboratories are more akin to moving targets, changing according to the talents and interests of the people available, the perceived interests of the granting agencies, and the newer developments in instrumentation.

Government laboratories, on the other hand, have no difficulty in retaining trained personnel, at least as long as the funding is continued. It is not surprising, therefore, that government laboratories tend to do well on intercalibrations, and to retain that capability over considerable periods.

2.1 Hydrocarbons

A priori, it would have been expected that the greatest drive towards intercalibration should have occurred in those areas most important to industry, such as petroleum hydrocarbon and organochlorine compounds in seawater, sediments, and marine organisms. Much has been done in this area (see, e.g. [5–15]), but the topic is covered in greater detail in the chapter by Zitko in this volume. The combination of the availability of reference materials and the participation in intercalibration exercises has resulted in a great improvement in both the accuracy and precision of these analyses.

In the case of the polychlorinated biphenyls (PCBs), one of the major improvements in analysis has resulted from the availability of pure reference compounds, so that concentrations need not be expressed as equivalents of some commercially available mixture of PCBs. While the techniques required to run these analyses are not simple, modern machine methods should put them within the reach of any laboratory which has the proper equipment and is willing to put in the time and effort.

2.2 Trace Metals

The analysis of trace metals in seawater, sediments, and marine organisms has always been a major preoccupation of marine chemists; the analytical literature contains methods for almost every element in the periodic table. Most of these methods have never been adopted by anyone other than the original author, as improvements in machine methods overtook methods based on chemical manipulations. In my own case, my Ph.D. thesis [16] described a number of novel methods for trace-metal analysis in sediments, all of which became irrelevant with the advent of atomic absorption analysis. The newer machine methods pin-pointed the wide range of values found in the literature for many trace metals, and by the late 1960s it was obvious that some sort of intercomparison or intercalibration was needed for the trace metals most commonly measured.

The Bedford Institute of Oceanography sponsored a pilot intercalibration study for trace metals in freshwater and seawater [17]. The results were somewhat disappointing, with some elements displaying a range of orders of magnitude. Since then, other intercalibration experiments have been run; the International Council for Exploration of the Seas (ICES), among others, has carried out a continuing series of trace-metal intercalibrations in sediments

[18–23], suspended particulate matter [24, 25], seawater [26–32], and biological materials [33, 34]. Unfortunately, the ICES reports are not carried by most libraries; they can be purchased directly from ICES in Copenhagen (<url>ices.info@ices.dk</url>).

The range of trace-metal concentrations reported in these intercalibrations has decreased, from orders of magnitude to factors of two or three. These ranges are still larger than is preferable, and it is obvious that some laboratories cannot determine certain metals reliably at seawater concentrations, but at least they are now aware of the fact, and can work to remedy the situation. In general, it is the laboratories new to the intercomparison studies that display the greatest deviation from the mean values.

2.3 Organic Carbon

Since 1964, the wet oxidation method of Menzel and Vaccaro [35] has been considered the standard method for dissolved organic carbon (DOC). High temperature oxidation methods have been developed [36, 37] and have worked successfully in freshwater, but the high carbonate and low DOC content of seawater has made the seawater technique much less dependable. While it seemed likely that some portion of the DOC present in seawater might require stronger oxidation than that supplied by persulfate in the presence of an excess of chloride ion, the demand for such analyses did not seem pressing enough to interest instrument companies in making DOC analyzers of the necessary sensitivity, and with the capability to withstand repeated injections of very hot seawater.

An intercalibration exercise for DOC in fresh and seawater samples involving fifteen European laboratories was reported in 1985 [38]. A variety of instruments and techniques was used, and a coefficient of variation of 5% was reported. The major cause of wild values was thought to be contamination.

In the mid-1980s a high temperature catalytic oxidation instrument was built in Japan, and applied to the analysis of dissolved organic nitrogen (DON) and DOC in seawater [39, 40]. The resultant high values for both DOC and DON, as well as an inverse correlation between DOC and apparent oxygen utilization (AOU), suggested that the wet oxidation method might have missed a large part of the pool of DOC and DON, particularly in deep water.

At a meeting held in Honolulu in January 1990, on the progress and future of marine organic geochemistry [41], it quickly became obvious that the most pressing problem in marine organic geochemistry was the reliability of the standard DOC methodology. If we could not keep track of the total organic carbon (TOC), there was no way we could be sure we were not missing large fractions which were difficult to oxidize. Accordingly, one of the major recommendations of the workshop was to conduct a round-robin intercomparison between the laboratories using various DOC and DON techniques, and to follow this intercomparison with a workshop on DOC and DON methodology.

This workshop was held in Seattle, WA, USA, in July, 1991, and the results of the workshop and intercomparison were published in a special issue of *Marine Chemistry* [42]. As is usually the case in initial round-robin intercomparisons, there were a few very high values which were probably due either to contamination or to an insensitive methodology, but most of the results, whether analyzed by high temperature catalytic oxidation or by wet oxidation, were within a fairly narrow band, covering a range of perhaps 2×, really rather good when compared with early intercomparisons for trace-metal analyses. The results tended to look even better when it was realized that a major problem in the calculation of DOC concentrations was the correction for the blank; organic carbon-free water is needed in order to determine a true machine blank. The results and statistics of the intercomparison are presented in a summary paper [43].

The workshop brought forward recommendations on techniques, including the proper calculation of machine blanks, sample collection and storage, and requirements for accuracy and precision for both DOC and DON analyses [44–47]. As a result of discussions held at this workshop, Suzuki recalculated some earlier published results, getting somewhat lower values [48], and rebuilt his DOC analyzer following suggestions from various colleagues [49]. The high values for DON found by Suzuki et al. [39] and by other workers [50] have not been found since; while there were too few results in the intercalibration study of DON to draw strong conclusions, more recent work [51, 52] suggests there is no large pool of undiscovered DON in the oceans, and that the usual persulfate oxidation releases essentially all of the nitrogen from the DON.

In this case, an intercalibration study served not only to test the ability of the various laboratories to perform an analysis, it also pointed out the steps in the determination which required further research. The high temperature catalytic oxidation (HTCO) technique has become much more common in this field, with several companies now supplying suitable instrumentation. This technique is not only simpler to run than the older wet oxidation methods, it also allows almost-real-time analyses at sea. The problem of the production of a proper blank, and therefore of a proper reference material for the DOC analysis, has yet to be solved; at this time it is being patched by the use of low DOC deep water as a blank, or of water which has passed through the combustion chamber of the analyzer and subsequently condensed. Placing DOC and total nitrogen (TN) detectors in-line has made simultaneous determinations of both quantities on the same sample a routine procedure [53].

Other organic components of seawater have also been the subject of intercalibrations. Two laboratories compared their methods of analysis for dimethyl sulfide in water [54]. The values found for seawater samples were comparable, but heavy cultures of phytoplankton produced differences which may have been related to the physiological state of the organisms and the methods used in sample preparation.

A subject of increasing importance has been the composition of oceanic particulate matter and the sediments derived from it. While the concentrations of organic carbon and nitrogen are much lower per unit volume than in the dissolved phase, the ability to concentrate the particle load by some form of filtration or centrifugation should result in the production of samples within the range of commercial CHN analyzers. Accordingly, we would expect a better performance from laboratories in an intercomparison test for particulate organic carbon (POC) and nitrogen (PON) than for DOC and DON. In an exercise involving ten laboratories [55], the standard error of the mean for organic C and N in sediments was 3%; in particulate matter, it was 7% for N, and 8% for C. The influence of the intercalibration test in the discussion in this paper is clear to see. In the 1970s, we would have been overjoyed with such results; in the 1990s, the authors call for better standard materials and more intercomparisons.

2.4 Miscellaneous Intercalibrations

Perhaps the most complete set of intercalibrations is the series for nutrients in seawater carried out by the ICES. In the fifth such exercise [56], results were submitted by 132 laboratories. The nutrients determined were nitrate, nitrite, ammonia, and phosphate. In addition, the laboratories were asked to submit any other determinands run on a routine basis. Eight laboratories submitted total nitrogen results, and six results for total phosphate. In general, nitrate and phosphate were determined well, while nitrite and ammonia showed a greater range of results. As in other intercalibration exercises, the newcomers to the exercise did not fare as well as the old hands. This intercalibration was held in conjunction with the Quality Assurance of Information for Marine Environmental Monitoring (QUASIMEME), a quality assurance initiative funded by the European Commission. This topic will be discussed later.

The degree of scatter in the results is surprising and a bit dismaying; these are the analyses run by most marine laboratories, in some cases for as long as the memory of any living oceanographer. In many laboratories, the analyses are so routine that they are run on some form of autoanalyzer. I would have expected much tighter distributions of results, even from first-time participants in the exercises.

ICES has also held an intercalibration exercise on the analysis of fatty acids in *Artemia* and of a feedstock used in aquaculture [57]. Eleven laboratories took part in the exercise. The results showed good within-lab precision for total lipids (3.6% CV for the feed, 4% for the *Artemia* nauplii), but inter-laboratory accuracy was not quite as good (5.2 and 8.7%, respectively). The results for individual fatty acids were not nearly as good. However, the results were still better than those found in an earlier intercomparison [58].

As part of the Joint Global Ocean Flux Study (JGOFS), a series of three intercalibration exercises for high performance liquid chromatography (HPLC) analysis of algal pigments was held. Most of the eight laboratories involved were good at the spectrophotometric determination of the pigments, but the HPLC results were not as good [59]. In this case, as in all of the JGOFS "core" determinations, recommendations on standard methods were made. These standard methods can be found at <url>http://ads.smr.uib.no/jgofs/publications/reports.htm</url>.

In light of the professed aims of the JGOFS program, perhaps the most important intercalibration experiment was the intercomparison of CO_2 measurements, in which twelve laboratories took part [60]. Two major methods of analysis were compared: a potentiometric measurement and a vacuum extraction method with manometric measurement. The various laboratories each displayed excellent precision, but poor agreement between them. In particular, the potentiometric method gave higher results than the extraction method. This research pointed out the need for better reference materials for this particular determination, to give laboratories a fixed point for their calibrations.

Since the measurement of CO_2 was so integral a part of the JGOFS research plan, and since the cruise plans could not await the prefect solution to this problem, an interim solution was devised for the US Indian Ocean CO_2 Survey of 1994–1996 [61]. Two coulometric total CO_2 analysis systems were placed on board the survey vessel at the start of the voyage and left there to be used by the several teams of analysts. Precision and accuracy were checked frequently by the use of certified reference materials, in this case seawater poisoned with mercuric chloride. The CO_2 content of the seawater had been measured by Keeling et al. using vacuum extraction and manometry. The authors were aware that any change in the geometry of the system could result in changes in calibration, and were careful to specify the conditions for care and feeding of the instrument.

It should be noted that this was not a case of intercalibration of instruments, but rather an avoidance of intercalibration by careful matching of two instruments which were then used for the whole series of cruises. The integrity of the data produced depended on the availability of the certified reference materials, which supplied fixed points on the calibration curves. In this particular case, three such fixed points were available, ranging in concentration from 1977 to 1993 µmol; in theory, these points are perhaps closer together than might ideally be preferable. In practice, the CRM values were close to those of the seawater samples, leaving no reason to doubt the experimental values.

2.5 Conclusions

Whatever the substance measured, the results of any first round-robin intercomparison suggest we do not measure as well as we think we do, at least at first. Participation in more intercomparisons seems to lead to better performance, hopefully as a result of greater care applied to the routine of analysis. There is always a danger, however, that samples known to be involved in an intercalibration run will receive a level of treatment not accorded to run-of-the-mill samples. We thus end up with faulty information; we do not want to know how well we *can* run samples, but how well we *do* run them, on an every-day basis.

My own introduction to intercalibration tests came as a senior laboratory technician in a US Army Air Corps hospital during World War II. About every six months we received sets of samples to be included in our daily analytical runs. It was part of my job to see that these samples were treated exactly like all the regular samples we were running, a job that required a high degree of low animal cunning. I never succeeded in getting them through without there being at least a suspicion, engendered by the degree of difficulty of a few of the samples.

All approved hospital and medical laboratories in the United States and most of those in Canada subscribe to the College of American Pathologists (CAP) surveys. The CAP sends out samples three or four times a year; we cannot hope for this frequency for marine samples, since at this time the people running the intercomparisons and intercalibrations are also full-time scientists. However, marine scientists are gathering together in larger groups as the tasks they are given grow in space and time. Perhaps in the not too distant future some group such as QUASIMEME in Europe or CORE in North America will have sufficient mass to afford a branch devoted to intercalibrations, and will be able to furnish samples at least yearly for intercalibrations for the substances most commonly determined.

The answer to our first question, can we measure what is given to us in a bottle?, must be a qualified "yes". Perhaps at first we cannot, but with the aid of certified reference materials and intercalibration experiments we can learn to do so, if we can afford the proper equipment. However, having once learned how, we cannot simply assume we now know and can do; we must continually demonstrate, to ourselves and to others, that we still are doing it right. Success at an intercalibration exercise describes only the present, and not any future state.

3
Methods Comparisons – Are We Measuring the Same Thing?

It is a truism of the analytical literature that whenever there is a multitude of methods for some substance in a particular matrix, the probability is high that none of these methods is really satisfactory. What is often not acknowledged is that these various methods may produce somewhat different answers, perhaps because of differences in interferences, or perhaps because different subcomponents of the substance are available to the methods. When a new analytical method appears in the literature, its limitations and constraints have seldom been explored exhaustively. This is particularly true if the method involves a new physical measurement and is monitored by a computer which relieves us of the tedium of calibration and calculation.

Before I am accused of being a Luddite, I would hasten to say that I have calculated 300-item regressions on a crank-powered Marchant calculator, and I would not care to return to that era; however, it is important that with any "black-box" analytical device the analyst not only understands the principles of its operation, but also the assumptions made by the programmer in setting up the calculations converting the actual measurement to the data entered into the notebook. The further experimenters get from the original data, the harder it is to see the clues which tell them their measurements are going astray. This becomes particularly important when the processes involved are inherently nonlinear.

As the intercalibration results have shown, even very experienced laboratories running well-accepted standard methods can produce unacceptably variable results on occasion; however, if reputable laboratories using different methods consistently produce different values for a given analysand, it is time to look into the methodology.

3.1 Dissolved Organic Carbon and Nitrogen

As an example, I will use the determinations of DOC and DON, partly because the discussions are very recent and partly because they are very familiar to me. The question of the accuracy of measurements of DOC in seawater has been disputed at least since the publications of Pütter [62] and Krogh [63]; the early work has been reviewed at length in an earlier publication [64]. While a variety of wet oxidation methods [65] were proposed for marine samples, the use of persulfate [35] provided the first real approach to a standard method. Persulfate oxidation, however, like all purely chemical oxidations, was a batch process, and not easily automated. A number of workers proposed photooxidation using ultraviolet (UV) light [66–70], and automated analytical systems which produced data almost in real time were soon constructed [71–73]. Commercial units soon appeared, but many of the units in the field were jerry-built, constructed out of parts scavenged from discarded autoanalyzers formerly used for nutrient analysis.

The photooxidation technique produced values of the same order as those produced by persulfate oxidation. There is no reason to expect that the two methods should be equally effective for all compounds to be found in seawater; it is surprising that so little comparative work has been done on these methods. I suspect the reasons for this are first, that both methods are effective oxidants for almost all of the pure compounds so far tested, and second, that few laboratories care to tool up for both of these methods. There are a fair number of intercomparisons between persulfate oxidation and high temperature catalytic oxidation, as a result of the controversy noted earlier, but very few direct comparisons of UV and persulfate; in one case [74], seawater samples run in parallel by both methods gave results some 15% lower by persulfate oxidation. While this might lead us to believe that UV oxidation decomposed some compounds not oxidized by persulfate, the possibility exists that the difference is due primarily to reactions between the sulfate radical and the chloride ion [75, 76], resulting in a lower efficiency of oxidation. The close correspondence between persulfate and UV oxidation and HTCO measurements on freshwater samples [77–79] reinforces this interpretation.

Intercomparisons between the persulfate and HTCO techniques showed considerable differences early in the investigations [36, 80], with later work [81–83] displaying much closer agreement between the methods. It is probable that some of the rapprochement results from a better understanding of the factors influencing the HTCO method, including choice and conditioning of the catalyst and the proper measurement of the instrument blank [84, 85].

Comparisons between the UV oxidation and HTCO methods produced slightly different results, in that the HTCO methods, at least in the early work, consistently produced slightly higher DOC values than did the UV methods [86–89]. These results are not too surprising, considering that two of the papers [86, 87] dealt with DOC in culture medium in which phytoplankton had grown, producing DOC values well above those normally found in seawater, and the other two papers similarly described special situations, in one case an estuary with a high colloidal content [88] and, in the other, the Marmara and the Black Seas [89].

If the difference between the two methods continues to exist as the HTCO method gains precision and accuracy through additional use, this could be a useful tool for the study of the composition of the DOC, particularly in axenic cultures. However, I do not expect comparative studies on DOC methods to continue; the HTCO analytical machines are so much simpler to operate, and so much faster at producing data, that I expect they will be doing the bulk of the DOC analyses in the world oceans. While there are still problems to be solved in the treatment of catalysts and in the preparation of standards and blanks [46, 90, 91], the persulfate method is not amenable to massive sample taking and the UV oxidation has its own problems, including ease of contamination of reagents and aging of the UV source.

The HTCO method offers the further advantage of relatively simple coupling of an NO detector, to give simultaneous DOC and total nitrogen (TN) values [92, 93]. Comparative studies between values of DON derived from wet oxidation and those from HTCO have shown no real difference between the methods [51, 93–96]. Ease of handling alone will probably ensure that most TN measurements are made by HTCO.

What has this exercise in comparative analysis shown us? I do not think we will ever know for sure whether there is a real difference between wet oxidation and HTCO DOC values for the open ocean. The advantages, both theoretical and practical, of the HTCO machines are so evident that their adoption as the standard method seems certain. The equivalence of the older wet oxidation data with the newer HTCO data will be accepted until such time as the total amount of newer data overwhelms the old. At the rate at which the HTCO data is accumulating, that time may only be a few years away.

3.2 Trace Metals

In the case of the trace metals, the first concern must be whether we are taking an uncontaminated sample. While the sample-handling rooms of the newer oceanographic vessels are very much improved over the rudimentary accommodation common at the middle of the century, the winches, cables, blocks, and the ship itself, are necessarily metal, and the utmost in precautions must be taken if the result is not to be a measure of the degree of contamination after sample taking. The sample preparation or analysis which is done on board is often performed under clean-room conditions, but if contamination has taken place in the early stages of sample handling [29, 97], no degree of further care can rescue the results.

Because of the extremely low concentrations of the trace metals in seawater, some method of concentration is commonly employed. Adsorption on Chelex 100 resin, chelation followed by extraction with an organic solvent, or co-precipitation, are techniques favored [98–100]. Intercomparison of concentration methods demonstrates that the choice of method depends on the metal chosen for analysis; no collection method appears to be clearly the best for all metals.

Once the metals have been concentrated, either voltammetric or graphite furnace atomic absorption methods can be applied.

What is then being measured is the total amount of the trace metal present in a form which passes through whatever type of screening material is serving as the upper limit for the "dissolved" phase. For many years this dividing line between "particulate" and "dissolved" was taken as 0.45 µm, although in practice it was often 1.2 µm or greater, depending on the filter chosen and on the treatment used to remove organic carbon from that filter. Recent developments in crossflow filtration, however, suggest that much of the trace metals and organic matter that we had regarded as in true solution is actually present in colloidal form [101–104].

Of equal interest to many geochemists is the speciation of the various trace metals; for the most part, this is approached through a wide variety of electrometric methods. These methods estimate the proportion of the trace metal bound by an organic ligand or, as shown by recent crossflow filtration work, bound to very small particles; the older work does not distinguish between these possibilities [105–113]. As is pointed out in other chapters in this volume (Kepkay, Zutic), even inclusion in the "colloidal" fraction is lumping together unlike species; the fate of trace metals which are part of the solid particles may be quite different from those which are gathered up by the loosely constructed bits and pieces which may ultimately constitute the gels.

It would appear, then, that we are at the point of making a major change in the way we measure trace metals in seawater; instead of a tripartite division into particulate, complexed, and ionic, we must add at least one compartment, and possibly two, for metals held in or on particles in the "colloidal" size range. The use of crossflow filtration is still far from a standard technique, and is not yet ready to be implemented in every oceanographic laboratory. However, the need has quite clearly been shown.

3.3 Large Particles

In general, samples of particulate matter taken from the water column are collected by filtration. A discussion of sampling and the problems involved can be found in the literature [114]. I suspect it came as no great surprise to anyone that changing the type or size of filter would result in changes in the amounts and sometimes in the nature of materials caught [115–117]. As we would expect, what was caught also depended on the degree of loading; at high particle loadings, as, for instance, in spring phytoplankton blooms, blockage of the filter openings led to decreased effective pore sizes, and better retention of small particles.

Again, a priori, we would not have expected to get comparable results from techniques involving such different physical principles as filtration, continuous-flow centrifugation, and deposition. While filtration depends on particle size and deposition upon particle density, continuous-flow centrifugation adds in the possibility of flocculation into one or a few large particles. The three methods appear to achieve roughly similar results [118, 119], although filtration does contain the possibility for adsorption of material which would otherwise be classed as "dissolved".

Continuous-flow centrifugation is attractive as a method for collecting very large amounts of particulate matter. Filtration is always limited by the problems involved in bringing large quantities of water on board, as well as by the lesser problems of clogging filters, leaking O-rings, and other misfortunes of the type visited upon Job. With centrifugation, the water can be pumped on board from the sampling device, passed through the centrifuge, and be sent back over the side. The major problems have been that the available continuous-flow units were designed for laboratory-size sample volumes, not oceans, and that such centrifuges are excellent gyroscopes, and require special mountings if they are to function in any real-world sea state.

It can be concluded, then, that the various ways of collecting particles could result in observable differences in the results. There is probably enough data now in the literature to test this assumption; it would require the separation of the data by collection method, area of ocean, depth, and season, and statistical testing of the resulting database. In the days of the crank-powered Marchant calculator, the statistics involved would have been formidable; today, the major effort would be in the building of the database. However, since cursory inspection of such data shows no large deviations with collection scheme, I doubt that the results would be startling enough to warrant the effort.

3.4 Conclusions

The answer to the question, "Does the method measure what we think it is measuring?", must also be a qualified yes, the qualification being that most methods when newly introduced appear to measure either something more or something different from what we will finally discover they really measure. This often results from an excess of enthusiasm on the part of the early users, a fault which I understand well and to which I am particularly prone. Before starting any large-scale investigation, ideally we should spend a period of time on the methods to be used, so that we understand just what our choice of methods is actually able to tell us. Unfortunately, the real world seldom offers us this kind of preparation time; the cruise date is always the day after tomorrow.

4 Are We All Sampling the Same Universe?

Once we are convinced that the various contributors to our large enterprise are really able to measure what they think they are measuring, we must next assume that the methods of sample collection they are using are such that, given comparable concentrations of the variables in question, they will produce comparable data. The proper way to test this assumption would be to run a multi-ship operation in which all of the ships involved were as close as possible, took their samples at the same time, each with its own variety of sampling gear, and with the samplers at the same apparent depth, and then exchanged samples throughout the fleet. Even then we would not expect perfect correspondence, for reasons discussed later, but we would be able to set realistic limits on between-ship variability. I have not heard of this being done in any systematic fashion; where this has been approached [120, 121], the results have shown a between-ships variability great enough to mask the effects sought. Until this sort of exercise is carried out, however, we do not know how to apportion any apparent variability we may see in the data.

There may be a number of sources for such variability, even given equal performance in intercalibration tests: the overall handling of samples may be different enough to influence the end results, even though the protocol used is apparently the same; the samplers used may be taking different pieces of the universe; or there may be local heterogeneities in the properties measured. The sample-handling problem does not need discussion; it is a constant problem, and can only be solved by good basic training of the personnel taking the samples and continuous watchfulness on the part of cruise leaders. Where possible, the sample-takers should also be the analysts, who thus have a stake in the accuracy of the results and understand the reasons for the rigid standards. An example of the kind of care necessary can be found in the discussion of CO_2 analyses for the JGOFS Indian Ocean cruises [61].

4.1 Differences Between Samplers

In the early stages of the development of a field-sampling program, the first aim is usually to discover whether there is anything out there we can measure, given the limits of the available instrumentation. A wide variety of sampling devices may be tried before the general opinion of the field settles on one or a few samplers. A good example of this sort of development can be seen in research into surface films.

4.1.1 *Surface Film Samplers*

Interest in the nature and quantity of material in the surface film in oceanic waters was fairly late in developing; the work of Baylor, Sutcliffe, and Hirschfeld [122] on the adsorption of phosphate onto bubbles brought these films to the attention of most chemical oceanographers. Shortly afterwards, the development of the Garrett screen [123] introduced a practical method for sampling the film. Early developments in this field are described in the literature [124].

Field workers soon found that the choice of screen material, as well as the thickness of the screen, could influence the amount of surface film material recovered. The Garrett screen was also a fair-weather collection method, since such collection had to be done from a small boat, away from the influence of the research vessel. A wide variety of devices and materials were soon tested and compared [125–128]. The general conclusion was that each device or sampling material collected a slightly different selection of compounds from the surface layer, so that direct comparisons between different collection protocols could not be made. The devices are sampling the same universe, but are taking different bites from it.

In the course of such investigations, workers in this area learned to what extremes they had to go to ensure that their sampling would not be contaminated by material emanating from the research vessel itself. Special sampling platforms had to be built [129], and extreme precautions used in sample taking and handling [130]. This last reference is to work which is certainly a milestone in the field; I do not expect to see it improved upon unless or until someone thinks of an exceedingly clever way to sample this material.

4.1.2
Water Samplers

The largest proportion of sampling of water properties involves taking water samples at known locations and depths. The samplers used are often adaptations of samplers already in use for many years, even though these samplers may be suspect as sources of contamination. Thus Nansen bottles survived well into the era of trace-metal and trace-organics analyses, and indeed still serve a useful purpose where salinity and temperature are the only data points required. However, as sampling requirements became more exacting, a succession of sampling bottles was developed, leading to the various low-contamination bottles now in general use. These bottles all do essentially the same job; they isolate a volume of water and enable it to be brought to the surface for various manipulations. They differ largely in the materials from which they are made, and in their closure mechanisms. There have been commercially available stainless steel samplers, designed for work on organic materials, and glass-lined samplers, for both trace metals and organics. How well they do their job is a matter of how much extraneous material they contribute to the sample of water, and in some cases, as in the collection of particulate material, whether the design of the sampler allows retention of some fraction of the particles [114].

When the substance being measured is present in ultra-trace quantities, as in some of the trace metals and many of the specific organic compounds, the volume of water which must be processed begins to get beyond the limit for safe handling aboard a rolling vessel. Each chief scientist has a bottle size above which they becomes distinctly uncomfortable; for me, this was around 30 l. I have been on cruises on which much larger samplers have been used, the biggest being a weather balloon, which could not be hoisted onto the ship. Instead, the contents of the balloon were pumped through a column which collected the required ingredient. This was definitely a fair-weather sampler.

An alternative to bringing up a large volume of water for processing would be to send the processing unit to the chosen depth, to do the processing in situ. One of the earliest units of this type [131] was designed for the collection of large quantities of particulate matter. While this unit suffered from the disadvantages of its large size and power requirements, so that only the largest oceanographic vessels could expect to handle it, subsequent models have incorporated local power supplies, permitting simultaneous operation of several units on the same cable, and provisions for in situ column extraction of specific dissolved components.

These units are still fairly large, and as more marine research institutes and departments become limited to estuarial and coastal research by budgetary restrictions, a need exists for such in situ pump sampling units which could be handled from a dugout canoe, if necessary. Such units have been designed and built [132–134] and are even offered commercially. To date, at least, they have not been as successful as might have been expected, largely because of problems in accurate measurement of the low flow rates needed, of leakage problems with extended use, and because of their relatively high initial cost. However, they offer the marine analyst the possibility of avoiding the problems of ultra-trace analysis by collecting the desired substance from a large volume of water. With the proper design of pretreatment columns, a good deal of separation chemistry can also be accomplished in situ [135, 136].

While I have seen no direct comparisons of sampling results from the small and large in situ pump samplers, I can think of no reason why their results would differ other than sheer scale; with the larger volume and higher pumping speed of the bigger samplers, a larger piece of the water column would be sampled. If there were heterogeneities of any magnitude involved, they might be included in the results of the larger pump sampler and not in those of the smaller unit.

4.1.3
Sampling Particulates

The standard method for collecting particulate matter has been to take water samples in the usual fashion, and to filter the water through a choice of filter. As I have already commented, the choice of filter will to some extent bias the results found. A larger bias can be introduced by the improper design of the sampling bottle. If a sampling bottle has its petcock located above the lowest point of the internal volume, any heavier particles will collect at this lowest point and not be included in the particulate load measured.

Another problem is the lack of representation of rarer particles in the volumes which can be sampled. In the course of some twenty years of work on particulate matter, I have examined thousands of filters under the microscope; I have yet to see my first foraminiferal test from samples taken below the surface layer, even in regions where the sediment was greater than 80% calcium carbonate. To relate what is happening at the sea surface to what has resulted at the sea floor, very much greater volumes of water must be sampled than is practical with water bottle samplers. This, of course, is what the in situ pumps were designed for in the first place, and what they do very well. The pumps give us a snapshot of what is in the water column at a particular time. With the newer pumps, which can be deployed at a series of depths simultaneously, the losses due to solution, remineralization, or predation in the distances between pump depths can be estimated, after some allowances have been made for the differences in background particulate burden in different water masses. What cannot be obtained from the pumps is any measurement of rate.

For measurements of rates of sedimentation, the instrument of choice has been the sediment trap. The sediment trap is engagingly simple in concept, but not all that easy to put into practice. When the requirement is to measure the amount of rainfall in a given region, a standardized bucket is positioned to catch what falls, and measure the fall at given intervals. The meteorologist can tell you immediately that it is much more complicated than that, and will discuss such matters as where to put the bucket, in what kind of surroundings, and what considerations must be given to wind direction and rain shadow, and many other factors it is unlikely anyone else would ever even think of. The same sorts of considerations are involved when a pan is deposited in the ocean for a given time, to see what will fall into it. Sediment trap experiments consist of setting collectors, ranging from relatively simple to very complex, into the ocean at various depths, either moored or free-floating, for time periods ranging from weeks to a year or more. At the end of the set time the traps are recovered and the catch examined and analyzed.

As already noted, sediment traps come in a variety of designs; it was recognized fairly early in their development that some sort of intercalibration would be necessary. Early work in the field suggested that the traps then in use gave essentially the same results [137], and that sediment trap flux data could be related to bottom sediment accumulation estimates [138]. However, work since then [139–151] suggests that the situation is much more complex, and that even small changes in trap design, as well as small hydrodynamic changes in the environment and differences in the methods of mooring, can have major effects on the amounts and kinds of materials caught.

It was recognized fairly early in the sediment trap work that biological materials left to accumulate in the traps for periods of up to a year would be subject to bacterial decomposition. If estimates of transfer of materials to the depths were to be made, some method of preservation of the accumulating material was required. A number of preservation methods of varying effectiveness have been described [152–161]. The method of preservation chosen must depend not only on its degree of effectiveness as a bactericide, but also on its effect on the analytical methodology to be employed. This is another case where the whole analytical procedure to be followed must be thought through before the sampling method is chosen.

The early work on sediment traps soon showed that not only the bacteria found the collected material interesting; after the addition of preservatives, in particular, larger organisms, presumably in search of a free lunch, the so-called "swimmers", were to be found in the traps [162–166]. The presence of one or a few large organisms can bias any calculation of material fluxes; therefore, ways were sought to exclude such organisms. To date, however, the usual method for handling swimmers is to pick them out of the collected material. There is no certain way to distinguish between an organism which entered the sediment trap as a swimmer looking for a free meal, and one which fell in as a legitimate large particle. The separation made by the pickers is, in the language of baseball, a judgment call. As a result, there is a built-in uncertainty in the flux calculations whenever swimmers are present in the samples. Also, any grid system built to exclude them will necessarily also serve to exclude particles of similar sizes.

There is also a matter I have never seen discussed, the problem of the effect of the presence of the sediment trap on the local productivity. When we wish to increase the fishing productivity in our coastal waters, a standard procedure is to build an artificial reef. When we bring in a mooring at the end of the sailing season, we are not surprised to find a healthy growth of marine life has made a home of the mooring. Why should we expect the presence of a mooring and a series of sediment traps, for example, to go unnoticed over a period of weeks to months? If a floating glass-ball net buoy can become a surface for goose barnacles to inhabit, should we not expect as much of a floating sediment trap? Any structure, fixed or floating, in marine waters soon carries its load of enhancers of the local productivity; the same must be true of sediment traps.

We should expect that the effect of such enhancement would be greater in highly productive waters than in less productive. The larger, more elaborate traps should show more effects than the small simple cylinders. I would also expect the effect to be nonlinear, coming to some equilibrium level with time as the useful surface area is filled.

4.2 The Heterogeneity of the Medium

The art of oceanographic sampling has improved greatly over the fifty years in which I have been an observer and participant. In my first years at sea, a wire angle was used and an amount of wire was let out to estimate the depth at which the sample was being taken; however, the location of where the sample had been taken was unknown until the thermometers, protected and unprotected, had been read and the calculations completed. Under these circumstances, comparisons between samples was usually done by assuming continuity between samples, drawing smooth curves against depth, and picking off values at standard depths.

With the introduction of the rosette sampler and the echo sounder, choosing the depth at which a sample is to be taken is a simple exercise in chart reading. Samples can be taken at any specified depth or in any specified water mass. Using much less sophisticated equipment on a cruise in the South Pacific [167], we were able to take water samples in the Antarctic Intermediate Water from 63 °S almost to the Equator, in order to demonstrate a horizontal continuity and vertical discontinuity in organic particulate content associated with a particular water mass.

Experience should have taught us that when our data is in conflict with continuity, it is probably continuity that is at fault; consider Gulf Stream eddies, "meddies", the coastal jets revealed by satellite photos, and salinity discontinuities in the surface layer. However, we are still faced with the problem that the oceans are so large and so deep, and our sampling patterns so sparse, that the concept of continuity must be invoked in order to make some sort of sense out of our data. Indeed, when we are considering large regions of the oceans, or whole water masses, it does make sense. The nutrients are low in the surface layers and higher at depth, and there are regions of the oceans which typically are higher in some components. It is when we are looking at smaller areas and closely spaced depths that continuity gets us into trouble.

In recent years, attempts have been made to investigate vertical and horizontal continuity by the construction of instruments permitting sampling at closer intervals in either space or time [168–176]. Employment of such instrumentation has shown us discontinuities on every scale examined [168, 169, 172, 177–179]. Understanding the causes and effects of such discontinuities is a major topic in itself, worthy of a separate review; in the context of this paper, the important point is the finding [177] that the patchiness encountered in the course of a 14-ship simultaneous sampling grid in the Baltic was below the resolution of a two-mile grid. This tells us that under the circumstances of this cruise, and possibly most cruises, there is a limit, set by the local patchiness of the water masses, where you cannot distinguish between analytical and sampling error and real differences in the universes sampled. Furthermore, I believe that in most cases we do not know what that limit is. For much of our work, as we deal with major differences between water masses, this limit is probably not important, but as we begin to study the details of behavior of marine ecosystems, using continuous recording devices [179], these limits will become important.

5 Conclusions

How, then, should we answer our three questions? The first, "How well can we analyze the sample once it is in our hands?", is fairly straightforward. For most of the components we wish to measure, we can find suitable methods. We can employ these methods with sufficient care, if we choose to make the effort. Our experience with intercalibration exercises teaches us that we do not always make the effort. A greater emphasis on quality assurance/quality control should become routine in all oceanographic laboratories, not just those involved in the big cooperative programs.

Does the method measure what we think it is measuring? I believe usually it does. When a new method, particularly of the black-box type, is introduced, there may be some question as to just what it is measuring, and how it arrives at its conclusions. As a group, however, we are skeptical enough to chase down discrepancies in results. If at times we seem to accept a new method without understanding exactly what it is measuring, as we discover its true limitations we tend to change our interpretations of its meanings.

The answers are somewhat different when we ask whether we are sampling the same universe. There are differences in water bottles – the choice would not be a stainless steel Nansen bottle with a brass valve if trace metals were being measured – but these are obvious sampling mistakes, not likely to occur, at least not more than once in a career. We would expect that filters with different physical structures, different pore sizes, and different composition would catch somewhat different distributions of particles, although the differences are not great [115]. Less obvious are the differences in sediment traps, for example, which show up in comparative experiments [148, 151]. The degree of difference even in composition of material sampled by the various surface film samplers [125, 127, 128] were unexpected.

The limiting differences, however, may well lie in the universes sampled. We do not know how far we can trust the principle of continuity. In the Baltic experiment on patchiness, we know the critical distance is less than two miles. This problem would most easily be solved for surface waters through satellite oceanography, providing we can get the resolving power necessary. What is really needed, however, is a shift from point sampling to continuous recording; whenever a continuous recording device has appeared for some new variable, discontinuities in the distribution of that variable have also appeared. Until we make the majority of our measurements in this manner, the most we can hope to do is to intercalibrate, compare instruments, assign differences to real differences in the universes sampled, and hope we are right.

6
References

1. Betzer PR, Pilson MEQ (1975) Deep Sea Res 22:117
2. Stoeppler M, Valenta, P Nürnberg, HW (1979) Z Anal Chem 297:22
3. Macdonald RW, O'Brien MC (1985) Anal Chim Acta 177:81
4. Waldichuk M, Jamieson WD, Berman SS (1987) Mar Pollut Bull 18:477
5. Farrington JW, Teal JM, Quinn JG, Wade T, Burns K (1973) Bull Environ Contam Toxicol 10:129
6. Medeiros GC, Farrington JW (1974) IDOE-5 intercalibration sample: results of analysis after sixteen months storage. In: NBS Spec Publ 409, Mar Pollut Monit (Petrol), p 167
7. Carlberg SR (1975/77) Ambio Spec Rept 5:269
8. Farrington JW, Teal JM, Medeiros GC, Burns K, Robinson EA Jr, Quinn JG, Wade TL (1976) Anal Chem 48:1711
9. Hilpert LR, May WE, Wise SA, Chester SN, Hertz HS (1978) Anal Chem 50:458
10. Law RJ, Portmann JE (1982) ICES Coop Res Rept 117:1
11. MacLeod WD Jr, Prohaska PG, Gennaro DD, Brown DW (1982) Anal Chem 54:386
12. Alford-Stevens AL, Budde WL, Bellar TA (1985) Anal Chem 57:2452
13. Uthe JF, Musial CJ (1986) ICES Coop Res Rept 136:81
14. de Boer J, Duinker JC, Calder JA, van der Meer J (1992) ICES Coop Res Rept 183:1
15. Reutergårdh L, Litzén K (1992) ICES Coop Res Rept 183:57
16. Wangersky PJ (1958) PhD thesis, Yale University
17. Macauley ID (1974) BIO Rept Ser BI-R-74-1:1
18. Ackermann JE, Bergmann H, Schleichert U (1979) Z Anal Chem 296:270
19. Brügmann L, Niemistö L (1987) ICES Coop Res Rept 147:1
20. Brügmann L, Niemistö L (1987) ICES Coop Res Rept 147:59
21. Jensen A (1987) ICES Coop Res Rept 147:51
22. Loring DH (coordinator) (1987) ICES Coop Res Rept 143
23. Loring DH, Rantala RTT (1988) Mar Chem 24:13
24. Yeats PA, Dalziel JA (1987) J Cons Int Explor Mer 43:272
25. Hovind H, Skei J (1992) ICES Coop Res Rept 184:1
26. Fukai R, Huynh-Ngoc L, Oregioni B, Morel A, Laumond F, Courau P, Nicolas E, Hardstedt-Romeo M, Molia R, Carrie M (1980) Thalassia Jugosl 16:171
27. Bewers JM, Dalziel J, Yeats PA, Barron JL (1981) Mar Chem 10:173
28. Alzieu C, Bewers JM, Duinker JC (1986) ICES Coop Res Rept 136:1
29. Bewers JM, Yeats PA, Westerlund S, Magnusson B, Schmidt D, Zehle H, Berman SS, Mykytiuk A, Duinker JC, Nolting RF, Smith RG, Windom HL (1986) ICES Coop Res Rept 136:5

30. Berman SS, Mykytiuk AF, Yeats PA, Bewers JM (1986) ICES Coop Res Rept 136:27
31. Cossa D, Courau P (1986) ICES Coop Res Rept 136:67
32. Topping G (1991) ICES Coop Res Rept 178:1
33. Brix H, Lyngby JE, Schierup H-H (1983) Mar Chem 12:69
34. Berman SS, Boyko VJ (1992) ICES Coop Res Rept 189:1
35. Menzel DW, Vaccaro RF (1964) Limnol Oceanogr 9:138
36. Sharp JH (1973) Mar Chem 1:211
37. Salonen K (1979) Limnol Oceanogr 24:177
38. Cauwet G (1985) Mitt geol-paläont Inst Univ Hamb 58:101
39. Suzuki Y, Sugimura Y, Itoh T (1985) Mar Chem 16:83
40. SugimuraY, Suzuki Y (1988) Mar Chem 24:105
41. Farrington JW (ed) (1992) Mar Chem 39
42. Hedges JI, Lee C (eds) (1993) Mar Chem 41
43. Hedges JI, Bergamaschi BA, Benner R (1993) Mar Chem 41:121
44. Williams PJleB, Bauer J (1993) Mar Chem 41:11
45. Hopkinson C, Cifuentes L (1993) Mar Chem 41:23
46. Sharp JH, Peltzer ET (1993) Mar Chem 41:37
47. Perdue EM, Mantoura F (1993) Mar Chem 41:51
48. Suzuki Y (1993) Mar Chem 41:287
49. Suzuki Y, Tanoue E, Ito H (1992) Deep Sea Res 35:185
50. Sagi T, Miyake Y, Saruhashi K. (1984) Bull Soc Sea Water Sci Jpn 38:353
51. Walsh TW (1989) Mar Chem 26:295
52. Williams PM, Bauer JE, Robertson KJ, Wolgast DM, Occelli ML (1993) Mar Chem 41:271
53. Álvarez-Salgado X, Miller AEJ (1998) Mar Chem 62:325
54. Turner SM, Malin G, Bågander LE, Leck C (1990) Mar Chem 29:47
55. King P, Kennedy H, Newton PP, Jickells TD, Brand T, Calvert S, Cauwet G, Etcheber H, Head B, Khripounoff A, Manighetti B, Miquel JC (1998) Mar Chem 60:203
56. Aminot A, Kirkwood D (1995) ICES Coop Res Rept 213:1
57. Coutteau P, Sorgeloos P (1995) ICES Coop Res Rept 211:1
58. Léger P, Bengtson DA, Sorgeloos P (1989) Analytical variation in the determination of the fatty acid composition of standard preparations of brine shrimp *Artemia*: an inter-laboratory exercise. In: Cowgill UM, Williams LR (eds) Hazard assessment, vol 12, ASTM STP 1027, American Society Testing Materials, Philadelphia, p 413
59. Latasa M, Bidigare RR, Ondrusek ME, Kennicutt MC II (1996) Mar Chem 51:315
60. Poisson A, Culkin F, Ridout P (1990) Deep Sea Res 37:1647
61. Johnson KM, Dickson AG, Eischeid G, Goyet C, Guenther P, Key RM, Millero FJ, Parkerson D, Sabine CL, Schottle RG, Wallace DWR, Wilke RJ, Winne CD (1998) Mar Chem 63:21
62. Pütter A (1909) Die Ernährung der Wassertiere und der Stoffhaushalt des Gewässer. Fischer, Jena
63. Krogh A (1934) Ecol Monogr 4:421
64. Wangersky PJ (1978) Production of dissolved organic matter. In: Kinne O (ed) Marine ecology, vol IV. Wiley New York, chap 4
65. Wangersky PJ (1993) Mar Chem 41:61
66. Beattie J, Bricker C, Garvin D (1961) Anal Chem 33:1890
67. Armstrong FAJ, Williams PM, Strickland JDH (1966) Nature 211:481
68. Ehrhardt M (1969) Deep Sea Res 16:393
69. Woelfel P, Sontheimer H (1974) Vom Wasser 43:315
70. Propp MV, Propp LN (1977) Okeanologiya 17:638
71. Collins KJ, Williams PJleB (1977) Mar Chem 5:123
72. Mueller H, Bandaranayake WM (1983) Mar Chem 12:59
73. Cauwet G (1984) Mar Chem 14:297
74. Gershey RM, MacKinnon MD, Williams PJleB, Moore RM (1979) Mar Chem 7:289
75. Peyton GR (1993) Mar Chem 41:91
76. McKenna JH, Doering PH (1995) Mar Chem 48:109

77. Kaplan LA (1992) Limnol Oceanogr 37:1119
78. Benner R, Hedges JI (1993) Mar Chem 41:161
79. Sharp JH, Suzuki Y, Munday WL (1993) Mar Chem 41:253
80. Powell RM, Bledsoe BE, Curtis GP, Johnson RL (1989) Environ Sci Technol 23:1246
81. Alperin MJ, Martens CS (1993) Mar Chem 41:135
82. de Baar HJW, Brussaard C, Hegeman J, Schijf J, Stoll MHC (1993) Mar Chem 41:145
83. Peltzer ET, Fry B, Doering PH, McKenna JH, Norrman B, Zweiful UL (1996) Mar Chem 54:85
84. Miller AEJ, Mantoura RFC, Preston MR (1993) Mar Chem 41:215
85. Sharp JH, Benner R, Bennett L, Carlson CA, Fitzwater SE, Peltzer ET, Tupas LM (1995) Mar Chem 48:91
86. Chen W, Wangersky PJ (1993) Mar Chem 41:167
87. Chen W, Wangersky PJ (1993) Mar Chem 42:95
88. Miller AEJ, Mantoura RFC, Suzuki Y, Preston MR (1993) Mar Chem 41:223
89. Tugrul S (1993) Mar Chem 41:265
90. Benner R, Strom M (1993) Mar Chem 41:153
91. Skoog A, Thomas D, Lara R, Richter K-U (1997) Mar Chem 56:39
92. Hansell, DA (1993) Mar Chem 41:195
93. Williams PM, Bauer JE, Robertson KJ, Wolgast DM, Occelli ML (1993) Mar Chem 41: 271
94. Maita Y, Yanada M (1990) Geochem J 24:245
95. Hedges JI, Bergamaschi BA, Benner R (1993) Mar Chem 41:121
96. Karl DM, Tien G, Dore J, Winn CD (1993) Mar Chem 41:203
97. Spencer MJ, Betzer PR, Piotrowicz SR (1982) Mar Chem 11:403
98. Brügmann L, Danielsson L-G, Magnusson B, Westerlund S (1983) Mar Chem 13:327
99. Boniforti R, Ferraroli R, Frigieri P, Heltai D, Queirazza G (1984) Anal Chim Acta 163:33
100. Bruland KW, Coale KH, Mart L (1985) Mar Chem 17:285
101. Greenamoyer JM, Moran SB (1996) Mar Chem 55:153
102. Helmers E (1996) Mar Chem 53:51
103. Reitmeyer R, Powell RT, Landing WM, Measures CI (1996) Mar Chem 55:75
104. Wen L-S, Stordal MC, Tang D, Gill GA, Santschi PH (1996) Mar Chem 55:129
105. Hering JG, Sunda WG, Ferguson RL, Morel FMM (1987) Mar Chem 20:299
106. Boussemart M, Benamou C, Richou M, Benaim JY (1989) Mar Chem 28:27
107. Donat JR, Bruland KW (1990) Mar Chem 28:301
108. Gledhill M, van den Berg CMG (1994) Mar Chem 47:41
109. Kerner M, Geisler C-D (1995) Mar Chem 51:133
110. Gordon AS, Dyer BJ, Kango RA, Donat JR (1996) Mar Chem 53:163
111. Midorikawa T, Tanoue E (1996) Mar Chem 52:157
112. Gledhill M, van den Berg CMG, Nolting RF, Timmermans KR (1998) Mar Chem 59:283
113. Witter AE, Luther GW III (1998) Mar Chem 61:241
114. Wangersky PJ (1994) Sampling and analysis of particulate and dissolved matter. In: Wotton RS (ed) The biology of particles in aquatic systems, 2nd ed. Lewis Publ, Boca Raton, chap 2
115. Wangersky PJ, Hincks AV (1980) Shipboard intercalibration of filters used in the measurement of particulate organic carbon. In: Albaiges J (ed) Analyical techniques in environmental chemistry. Pergamon Press, New York, p 53
116. Brzezinska-Paudyn A, Balicki MR, Van Loon JC (1985) Water Air Soil Pollut 24:339
117. Altabet MA (1990) Limnol Oceanogr 35:902
118. Etcheber H, Jouanneau JM (1980) Estuary Coast Mar Sci 11:701
119. Bates TS, Hamilton SE, Cline JD (1983) Estuary Coast Shelf Sci 16:107
120. Law RJ, Marchand M, Dahlmann G, Fileman, TW (1987) Mar Pollut Bull 18:486
121. Leppånen J-M, Kononen K, Behrends G, Hansen G (1990) Finn Mar Res 257:37
122. Baylor ER, Sutcliffe WH Jr, Hirschfeld DS (1962) Deep Sea Res 9:120
123. GarrettWD (1965) Limnol Oceanogr 10:602

124. Wangersky PJ (1976) Ann Rev Ecol Sys 7:161
125. Daumas RA, Laborde PL, Marty JC, Saliot A (1976) Limnol Oceanogr 21:319
126. Kjelleberg S, Stenstrom TA, Odham G (1979) Mar Biol 53:21
127. Van Vleet ES, Williams PM (1980) Limnol Oceanogr 25:764
128. Carlson DJ (1982) Mar Chem 11:189
129. Williams PM, Long DL, Price CC, Robertson KJ, Van Vleet ES (1982) Deep Sea Res 29: 641
130. Williams PM, Carlucci AF, Henrichs SM, Van Vleet ES, Horrigan SG, Reid FMH, Robertson KJ (1986) Mar Chem 19:17
131. Bishop JKB, Edmond JM (1976) J Mar Res 34:181
132. Johnson BD, Wangersky PJW (1984) NRCC Rept No 23609
133. Johnson BD, Wangersky PJ, Zhou X (1987) Mar Chem 22:353
134. Petrick G, Schulz-Bull DE, Martens V, Scholz K, Duinker JC (1996) Mar Chem 54:97
135. Slauenwhite DE, Johnson BD, Wangersky PJ (1986) NRCC Rept No. 26325
136. Slauenwhite DE, Wangersky PJ (1996) Mar Chem 54:107
137. Dymond J, Fischer K, Clauson M, Cobler R, Gardner W, Richardson MJ, Berger W, Soutar A, Dunbar R (1981) Earth Planet Sci Lett 53:409
138. Bruland KW, Franks RP, Landing WM, Soutar A (1981) Earth Planet Sci Lett 53:400
139. Staresinic N, Rowe GT, Shaughnessy D, Williams AJ III (1978) Limnol Oceanogr 23: 559
140. Blomqvist S, Håkanson L (1981) Arch Hydrobiol 91:101
141. Blomqvist S, Kofoed C (1981) Limnol Oceanogr 26:585
142. Staresinic N, von Bröckel K, Smodlaka N, Clifford CH (1982) J Mar Res 40:273
143. Gardner WD (1985) Deep Sea Res 32:349
144. Butman CA, Grant WD, Stolzenbach KD (1986) J Mar Res 44:601
145. Butman CA (1986) J Mar Res 44:645
146. Noriki S, Tsunogai S (1986) J Oceanogr Soc Jpn 42:119
147. Asper VL (1987) Mar Technol Soc J 21:18
148. Baker ET, Milburn HB, Tennant, DA (1988) J Mar Res 46:573
149. Honjo S, Spencer DW, Gardner WD (1992) Deep Sea Res 39:333
150. Gust G, Byrne RH, Bernstein RE, Betzer PR, Bowles W (1992) Deep Sea Res 39:1071
151. Gust G, Michaels AF, Johnson R, Deuser WG, Bowles W (1994) Deep Sea Res 41:831
152. Gardner WD, Hinga KR, Marra J (1983) J Mar Res 41:195
153. Knauer GA, Karl DM, Martin JH, Hunter CN (1984) J Mar Res 42:445
154. Deming JW (1985) Mar Ecol Prog Ser 25:305
155. Ducklow HW, Hill SM, Gardner WD (1985) Continent Shelf Res 4:445
156. Lee C, McKenzie JA, Sturm M (1987) Limnol Oceanogr 32:83
157. Banse K (1990) Deep Sea Res 37:1177
158. Gundersen K, Wassmann P (1990) Mar Ecol Prog Ser 64:187
159. Buesseler KO (1991) Nature 313:420
160. Lee C, Hedges JI, Wakeham SG, Zhu N (1992) Limnol Oceanogr 37:117
161. Khripounoff A, Crassous P (1994) Deep Sea Res 41:821
162. Harbison GR, Gilmer RW (1986) Deep Sea Res 33:1017
163. Coale KH (1990) Limnol Oceanogr 35:1376
164. Michaels AF, Silver MW, Gowing MM, Knauer GA (1990) Deep Sea Res 37:1285
165. Peterson W, Dam HG (1990) Limnol Oceanogr 35:448
166. Steinberg DK, Pilskaln CH, Silver MW (1998) Mar Ecol Prog Ser 164:157
167. Wangersky PJ (1976) Deep Sea Res 23:457
168. Wangersky PJ (1974) Limnol Oceanogr 19:980
169. Wangersky PJ (1978) Int Rev Ges Hydrobiol 63:567
170. Anderson JJ, Okubo A (1982) Deep Sea Res 29:1013
171. Mackas DL, Owen RW (1982) Deep Sea Res 29:883
172. Johnson BD, Wangersky PJ (1985) Deep Sea Res 32:1143
173. Dunn J, Hall CD, Heath MR, Mitchell RB, Ritchie BJ (1993) Deep Sea Res 40:867
174. Schüssler U, Kremling K (1993) Deep Sea Res 40:257

175. Widder EA, Case JF, Bernstein SA, MacIntyre S, Lowenstine MR, Bowlby MR, Cook DP (1993) Deep Sea Res 40:607
176. Daniel A, Birot D, Blain S, Tréguer P, Leïldé B, Menut E (1995) Mar Chem 51:67
177. Lehman JT, Scavia D (1982) Science 216:729
178. Kahru M, Leppånen J-M, Nômmann S, Passow U, Postel L, Schulz S (1990) Mar Ecol Prog Ser 66:301
179. Taylor CD, Howes BL (1994) Mar Ecol Prog Ser 108:193

CHAPTER 8

Lipid and Phenolic Biomarkers in Marine Ecosystems: Analysis and Applications

C. C. Parrish[1], T. A. Abrajano[2], S. M. Budge[3], R. J. Helleur[4], E. D. Hudson[4], K. Pulchan[5], C. Ramos[6]

[1] Ocean Sciences Centre, Memorial University of Newfoundland, St. John's, Newfoundland, A1C 5S7, Canada
E-mail: cparrish@morgan.ucs.mun.ca

[2] Department of Earth and Environmental Sciences, Science Center, Room 1C25, Rensselaer Polytechnic Institute, 110 8th Street, Troy, NY 12180-3590 USA
E-mail: abrajt@rpi.edu

[3] Biology Department, Dalhousie University, Halifax, Nova Scotia, B3H 4J1, Canada
E-mail: budges@is.dal.ca

[4] Department of Chemistry, Memorial University of Newfoundland, St. John's, Newfoundland, A1B 3X7, Canada
E-mail: rhelleur@morgan.ucs.mun.ca

[5] Department of Earth Sciences, Memorial University of Newfoundland, St. John's, Newfoundland, A1B 3X7, Canada

[6] Institute of Chemistry, University of the Philippines, Diliman, Quezon City 1101, Philippines

Biomarkers are compounds or groups of compounds that can be used as signatures of individual organisms or groups of organisms, or of certain environmental processes. Lipid and phenolic biomarkers can be used to assess the health of an ecosystem and the degree to which it has been influenced by terrestrial and anthropogenic inputs. Lipid classes and fatty acids can be used to determine production of marine biogenic material of dietary value to pelagic and benthic organisms. Polycyclic aromatic hydrocarbons and 5β-stanols such as coprostanol can be used to determine pollutant loading from oil spillage or sewage and the phenanthrene/methylphenanthrene ratio can be used specifically as an indicator of wood burning. *N*-alkanes and thermochemolysis products in cores can show the sensitivity of sediments to changes in land use patterns near the land margin. The relationship between marine and terrestrially derived products in sediment cores can be used to indicate the degree to which land use changes have impacted the pattern of marine biogenic productivity in the area. Stable isotope and multivariate analyses are particularly useful for biomarker validation.

Key words: Molecular signatures, Stable isotopes, Chemometrics.

The Handbook of Environmental Chemistry Vol. 5 Part D
Marine Chemistry
(ed. by P. Wangersky)

1 Introduction

Molecular biomarkers are easily determined compounds that tell us about the history of a sample. They can be signatures of the condition of a sample, they can tell us about past events and even about the future when certain compounds are used as early warning signals. Both molecular and isotopic analyses of biomarkers have been extensively used in geochemical studies [e.g. 1, 2] but there is now increasing interest in their use in ecological studies. Indeed, further biomarker studies of modern environments will greatly aid source identification in sediments [e.g. 3].

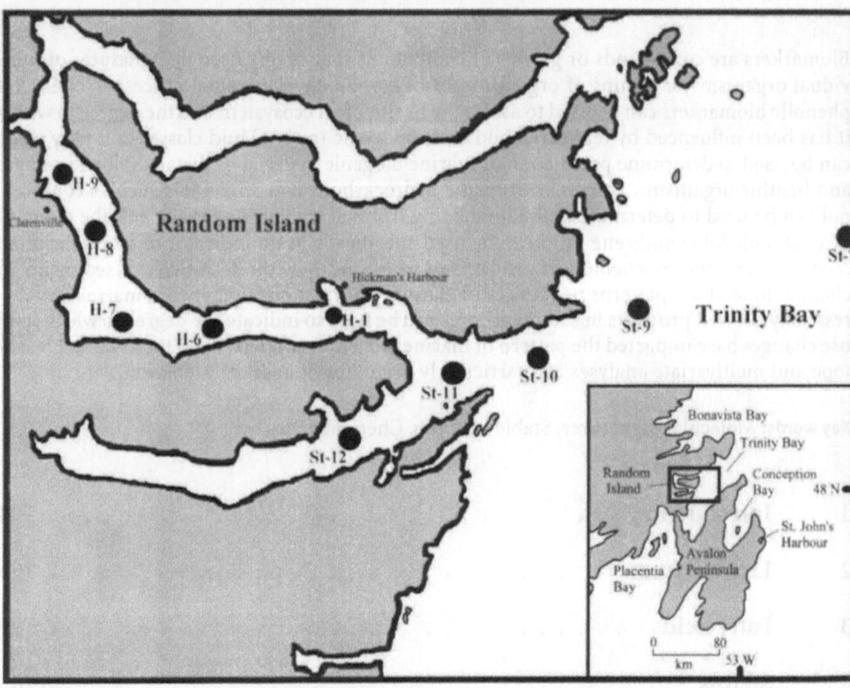

Fig. 1. Station locations in Trinity Bay and around Random Island. Trinity Bay is a large fjord-like bay which historically had an abundance of spawning cod during spring. It is about 100 km long and 30 km wide, with a maximum depth of about 590 m and is representative of several cold ocean bays around Newfoundland

A marine biomarker can be a DNA fragment or an enzyme that shows us that a fish or an invertebrate has been exposed to a xenobiotic [e.g. 4–7], but our focus here will be on smaller molecules that are determined using standard chromatographic techniques. Lipids are one such group that are receiving increasing amounts of attention in ecological [8] and biogeochemical [9, 10] studies. Less well known is the biomarker potential of phenolic compounds determined by pyrolysis gas chromatography. In this chapter we will compare and contrast the use of these two types of biomarkers using research we undertook in Trinity Bay, Newfoundland as a case study (Fig. 1). The use of biomarkers to investigate the marine ecosystem in the cold waters off the coast of Newfoundland is new and timely given current problems with the ground fishery [11, 12] and the increasing interest in aquaculture and oil field development in the area. This is the first study to compare different lipids and phenolics in plankton, settling particulate material and sediments in a cold fjord-like environment. Polycyclic aromatic hydrocarbons and fecal sterol biomarkers were used to evaluate pollutant loading. In addition, the PAH distribution patterns were determined in order to fingerprint anthropogenic sources as being petrogenic or pyrolytic. Total lipids, lipid classes, polyunsaturated fatty acids and sterols were used to examine marine inputs, as were total fatty acids derived from thermochemolytic analyses. The ΣnC_{26}-nC_{35}/ΣnC_{16}-nC_{25} *n*-alkane ratio and phenolic compounds were used for terrestrial plant inputs to the marine ecosystem. Finally, 3,4-dimethoxybenzoic acid methyl ester was used as a specific marker for wood inputs.

2 Lipid Classes

Lipids are carbon-rich compounds with a very high energy value, making them important fuels in marine ecosystems. Marine lipids are usually extracted using some kind of simplification of the Folch et al. [13] procedure in which the sample is ground in chloroform and methanol (2:1). Such an extract may contain as many as 16 different subclasses of both biogenic and anthropogenic origin [14]. The heterogeneous nature of lipids means that much information can be gained by determining individual classes and in the process, lipids can be separated from non-lipid contaminants.

Lipid classes can be separated by thin layer chromatography on silica gel coated Chromarods [15] or plates [16, 17]. Both rod and plate TLC are amenable to quantitation: plates can be scanned in a densitometer, while Chromarods can be passed through the flame ionization detector of an Iatroscan. An advantage of the Chromarod-Iatroscan TLC/FID system is the partial scanning facility that permits extensive analysis of a single sample on a single rod. By separating out all the lipid classes in this way, much greater confidence in the identities of individual peaks is obtained and any non-lipid material remains at the origin. In addition, the Chromarod-Iatroscan procedure is sensitive, with a detection limit of around 50 ng, and it has been successfully used in an intercalibration exercise between laboratories in different countries [18]

Marine samples with high proportions of polyunsaturated fatty acids may present problems with identification and quantification by TLC/FID. With

Chromarod separations, peak splitting may occasionally be observed in the wax ester or triacylglycerol region of chromatograms. This is due to the presence of molecular species with widely differing degrees of unsaturation [19], and it could lead to misidentification. The second part of a split wax ester peak could be mistaken for a fatty acid methyl ester and the second part of a split triacylglycerol peak could be mistaken for a free fatty acid. There may also be a problem if Iatroscan calibration has been done with saturated standards. This could cause an 18–70% underestimation of the amount of a lipid class present [20, 21] because unsaturated lipids give a lower detector response and because of the broadening of Chromarod bands containing significant proportions of polyunsaturated molecular species. Band broadening also results in lower responses [22, 23]. However, the difference in absolute response due to unsaturation is quite small [19] and the effect of band broadening can be counteracted by exposing the band to multiple developments in schemes that involve solvent focusing, double developments, or partial scanning and redevelopment. The fact that Iatroscan values for aquatic samples are routinely 80–95% of those obtained by gravimetry and other methods [18, 24–28] attests to the general applicability of saturated standards in calibrations.

Another approach to obtaining synoptic marine lipid class data is by short column gas chromatography [29]. In terms of the information provided, this profiling method is located between detailed fatty acid analyses and the class analyses provided by the Chromarod-Iatroscan system. It groups compounds according to carbon number within each class. By summing the groups of molecular species within each class, total class amounts can be obtained. This method is readily automated and has been successfully applied to a wide range of marine samples [29].

In marine ecosystem studies, two lipid classes that are of particular interest are the triacylglycerols and phospholipids which are biochemically related. They both possess a glycerol backbone to which 2 or 3 fatty acids are esterified and they also share a common precursor. Triacylglycerols are a very important energy storage substance and have been used as a condition index for marine fauna [30]. In our Trinity Bay study (Fig. 1) we found that in spring, input rates of total lipids increased to a level typical of a highly productive oceanic upwelling region [31]. At this time, the highest flux of triacylglycerol was observed in the lipid material falling through the water column (Fig. 2), indicating storage of energy in bloom organisms and its transfer to the benthos. The triacylglycerol fluxes measured at all 3 depths in late spring, together with some of those observed during summer, summer/fall, and early spring in Trinity Bay were higher than those reported for various oceanic regions, including the Peru upwelling region [31]. The high triacylglycerol fluxes in late spring are undoubtedly related to low nutrient concentrations: diatoms increase triacylglycerol synthesis when nitrate or silicate supply is low [32–34]. Triacylglycerols, together sometimes with steryl esters and free fatty acids, are also important in determining PCB concentrations in marine biota [35].

Phospholipids are essential components of membranes where they share a structural function with sterols. Phospholipids can be used to indicate freshly biosynthesized material [36] and individual ones may be used to distinguish

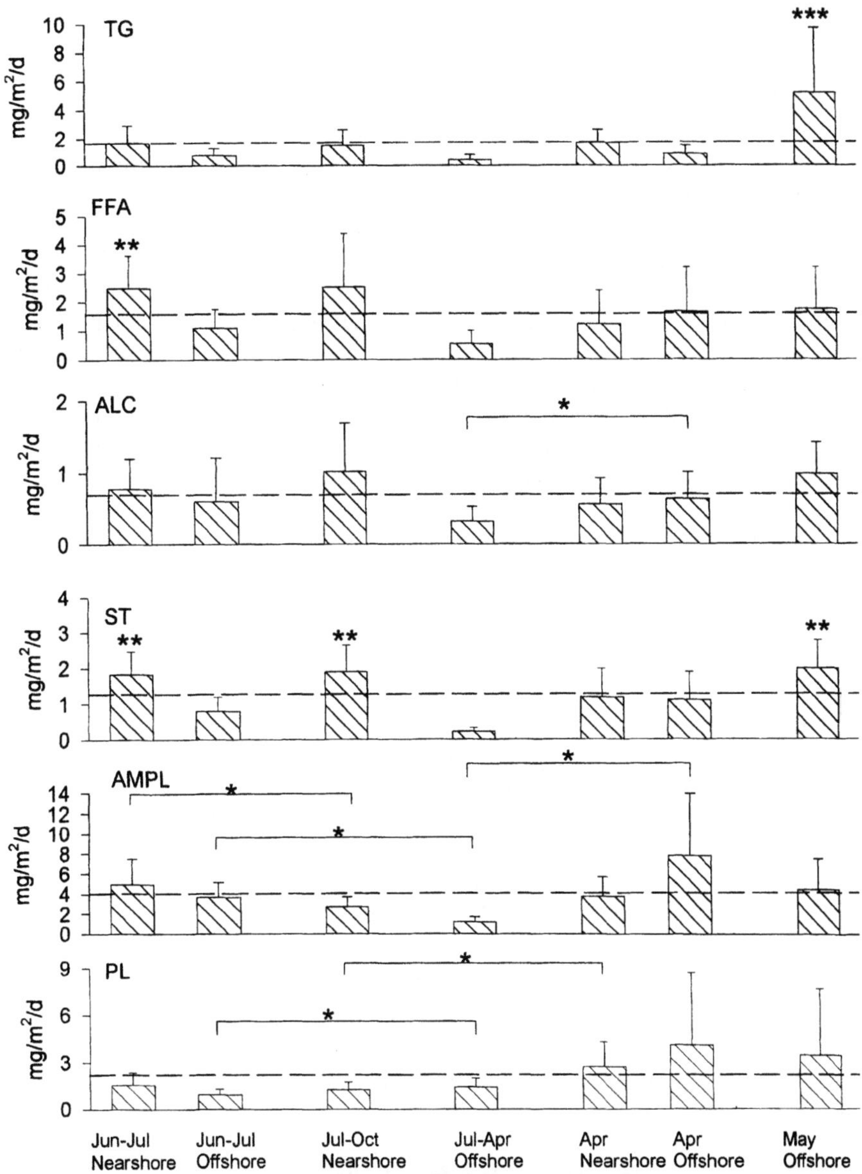

Fig. 2. Fluxes of particulate matter through the water column in Trinity Bay, Newfoundland. Rates are the averages of traps at 3 depths (50, 75, 100 m) for lipid class inputs. The *error bars* indicate one standard deviation above the mean; the *dashed line* is the annual mean flux. *TG*: triacylglycerol, *FFA*: free fatty acid, *ALC*: alcohol, *ST*: sterol, *AMPL*: acetone-mobile polar lipids, *PL*: phospholipid. $^{***}p < 0.05$: Significantly higher than the annual mean flux as well as all other fluxes. $^{**}p < 0.05$: Significantly higher than the annual mean flux. $^{*}p < 0.03$: Significantly different fluxes in different seasons

bacteria and phytoplankton [37, 38]. In Trinity Bay, the highest fluxes of phospholipids occurred in spring (Fig. 2) indicating massive transport of membrane material through the water column. The flux of acetone-mobile polar lipids was also highest at this time. This class is a mixture of pigments and chloroplast-associated glycolipids and thus strongly indicates plant material transiting the water column. The lowest fluxes of almost all lipid classes were found in the fall/winter samples.

Ketones are a class of lipids that were rarely seen in plankton and sediment trap samples from Trinity Bay, although there was a significant level in sediments. This suggests the possibility of their use as an internal standard in lipid analyses of water column samples. However, they are present in some prymnesiophytes and, in fact, there is much interest in their use as paleotemperature indicators [e.g. 39, 40], although methyl and ethyl ketone concentrations respond to nitrogen limitation as well [41]. Their rarity combined with the fact that they may comprise as much as a quarter of the lipids of some species [42], however, suggests that as a class, ketones could be excellent water column markers in some areas.

Lipid breakdown can be assessed from the free fatty acid content [14, 36, 43] or the free fatty acid plus alcohol content [44]. The proportions of free fatty acids were at their lowest during the spring bloom in Trinity Bay with values as low as 5%, reflecting a relative lack of acyl lipid degradation at that time. Fluxes of free fatty acids and free aliphatic alcohols, were highest in the summer/fall samples. The high alcohol flux suggests degradation of zooplankton-derived wax esters and microalgal chlorophylls was greatest then. However, free fatty acid levels and perhaps free alcohol levels may be easily overestimated. Recent papers suggest free fatty acid levels in diatoms should really be close to zero, and that problems are encountered at the sample collection and extraction stage [45, 46]. The use of boiling water to deactivate lipolytic enzymes is to be recommended.

An index to indicate the degree of breakdown has been suggested by Weeks et al. [47] which they term the hydrolysis index (HI). It is defined as:

$$HI = \frac{\text{(free fatty acids + alcohols)}}{\text{(total non-polar acyl lipids + products)}} \times 100$$

An alternative lipolysis index based on that of Weeks et al. [47] has also been proposed [44]:

$$LI\ (\%) = \frac{\text{(free fatty acids + free alcohols)}}{\text{(total acyl lipids + free alcohols)}} \times 100$$

This lipolysis index is usually strongly correlated with that of Weeks et al. However, it differs from theirs in that it takes into account all sources of hydrolysis products, polar and non-polar, since only hydrocarbons and sterols are missing in the denominator. In our Trinity Bay study, the lipolytic breakdown indices of net-tow samples were the same as in the sediment trap material suggesting plankton lipids were well preserved in traps.

Thus, in addition to overall caloric content, lipid class data can provide information about the condition of a sample in terms of age of material or nutrient limitation of algae. However, lipid class information is particularly valuable when used in conjunction with determinations of individual compounds.

3 Fatty Acids

Acyl lipids are almost always the major contributors to a marine lipid extract. The fatty acid moieties are often determined individually as fatty acid methyl esters (FAME) which are commonly produced from lipid extracts by transesterification using one of several techniques. We have found that a simple and rapid method employing 10% boron trifluoride in methanol at 85 °C [48] produces equivalent or superior yields to other acid-catalyzed techniques with no evidence of polyunsaturated fatty acid (PUFA) loss or artifact formation. FAMEs are most commonly analyzed using a gas chromatograph (GC) equipped with flame ionization detection. The use of a mass spectral detector is invaluable in determining the structure of unusual fatty acids. A polar GC column, such as one coated with a polyethylene glycol [49], is necessary for adequate separation of isomers and a gas-line oxygen scrubber is essential [50]. In this way 30–35 fatty acids can be routinely separated. They can be named using a convenient shorthand notation of the form A:BωX, where A represents the number of carbon atoms, B gives the number of double bonds and X gives the position of the double bond closest to the terminal methyl group. With this naming system, all double bonds are assumed to be methylene-interrupted and *cis* in configuration.

Using fatty acid biomarkers, general phytoplankton sources in marine samples may be determined. The ratios of 16:1/16:0 and the sum of all fatty acids having 16 carbon atoms to the sum of all fatty acids having 18 carbon atoms ($\Sigma C_{16}/\Sigma C_{18}$) were originally proposed by Claustre et al. [51] as diatom markers. In conjunction with this, Bodennec et al. [52] suggested that values of 16:1/16:0 greater than 1.6 could be interpreted as signalling the predominant presence of diatoms. Claustre et al. [51] interpreted an increase in values of both of these ratios as representative of increased proportions of diatoms. In addition to these two markers, the fatty acid, 16:4ω1, can be also used to assess the importance of diatoms. This acid is commonly found in diatoms [53, 54] but is very rarely encountered in other phytoplankton, and it can be used as a general marker for diatoms. In Trinity Bay, a detailed examination of the fatty acid composition of plankton samples collected during spring clearly showed the development of the spring diatom bloom with a maximum reached in all three markers in May (Fig. 3a). In addition to elevated amounts of 16 carbon fatty acids, diatoms produce large proportions of 20:5ω3. Dinoflagellates, on the other hand, generally contain higher proportions of 22:6ω3. The combination of those two fatty acids in a ratio, 22:6ω3/20:5ω3, produces a marker which reflects the predominance of dinoflagellates versus diatoms. It should be noted, however, that this marker only applies in environments such as Trinity Bay where dinoflagellates and diatoms are the major producers of 22:6ω3 and 20:5ω3, respectively

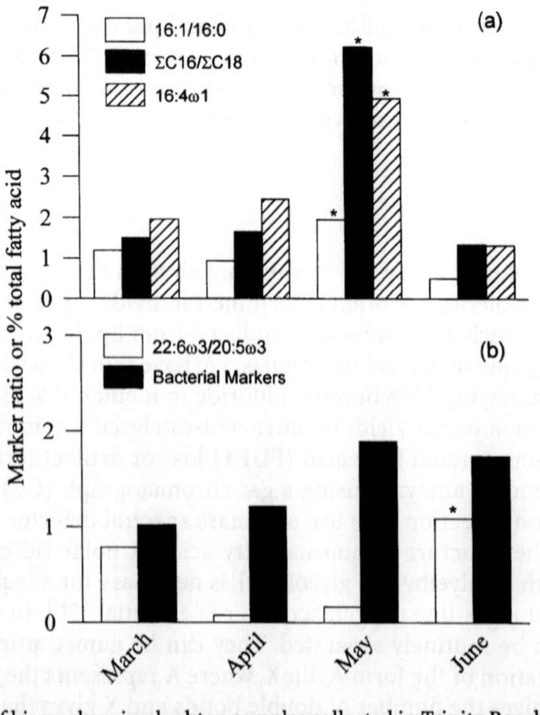

Fig. 3 a, b. Levels of biomarkers in plankton samples collected in Trinity Bay during the spring bloom of 1996 (After [50]). **a** Diatom markers and **b** dinoflagellate and bacterial markers. *Significantly different ($p < 0.05$) from the same marker value in all other months

[50]. In other environments, other microalgae may contribute significantly to the 22:6ω3 pool. In Trinity Bay, dinoflagellates were found to be a relatively more important source of fatty acid material before and after the spring bloom (Fig. 3b).

Several fatty acids, specifically 15:0, 17:0 and all branched fatty acids, are produced primarily by both aerobic and anaerobic bacteria [55–57] and the sum of those fatty acids has been used to estimate bacterial contributions [58–61]. A comparison of bacterial markers in plankton, sediment trap and sediment samples showed the lowest values, with little variation, in plankton samples (Fig. 3b), and the greatest bacterial levels in sediments. The sediment traps, containing partially degraded material, had bacterial marker levels intermediate between the other two sample types, and levels of bacterial markers increased with increasing period of deployment. However, there are conflicting theories concerning the usefulness of these markers and, for that reason, bacterial markers should only be employed with caution. For instance, in a recent paper, Harvey and Macko [57] did not find a correlation between total fatty acids attributed to bacteria and bacterial carbon, and they suggest that bacterial fatty acids only be used as qualitative tools to estimate bacterial contributions. Wakeham [62] also points out that fatty acids of common oceanic bacteria may not be compositionally different from planktonic fatty acids so that bacterial contributions may not be easily discernible. On the other hand, Canuel and Martens [63] propose that bacterial biomass in coastal sediments can be calculated from bacterial markers such as those used here. In light of these conflicting theories, it seems that bacterial fatty acids should only be used to determine bacterial levels relative to other samples in the same study.

In marine environments, there is always interest in determining terrestrial plant contributions. Long-chain (> 24 carbons) fatty acids are often used as terrestrial plant indicators [61, 64, 65] but their analysis is problematic as those fatty acids do not elute within the maximum temperature limits of most polar columns. Alternatively, the fatty acids, 18:2ω6 and 18:3ω3, found in elevated amounts in most terrestrial plants [49], may be used as terrestrial markers [50, 66]. By examining terrestrial plant, pollen, riverine and plankton samples, an arbitrary threshold of 2.5% has been assigned to this indicator [50]. In this way, samples with values above this may be considered to have terrestrial material as a significant source of organic matter. Preliminary studies with compound specific isotope analyses offered support for this marker as the analyses showed that 18:2ω6 and 18:3ω3 in riverine samples were the most depleted of all fatty acids with $\delta^{13}C$ values (see Sect. 7 below) of –33.4 and –33.6‰, respectively. In Trinity Bay, in spring and early summer, values of those markers were quite low in the plankton and sediment trap samples, but in late summer and fall, terrestrial markers reached a maximum in trap samples. Those markers also comprised almost 10% of total fatty acids in the sediments, suggesting that terrestrial material was preserved to a greater extent than marine fatty acids in sediments.

Zooplankton grazing is an important link between lower and higher trophic levels and fatty acid biomarkers may also be employed to determine the importance of zooplankton sources. Generally, herbivorous and omnivorous zooplankton feeding predominantly on phytoplankton contain elevated amounts of long-chain monounsaturated fatty acids within the wax ester lipid fraction [67–69]. Because of this, the sum of 20:1 and 22:1 fatty acids may be used as a zooplankton marker. In our study, zooplankton markers were highest in plankton samples, presumably because of the ample supply of phytoplankton as a food source during the spring bloom. These markers, however, also comprised approximately 6% of total fatty acids in both sediment traps and sediments, suggesting zooplankton-sourced material made a substantial contribution to the fatty acid pool. Over time, the zooplankton markers varied much like the terrestrial markers, with zooplankton becoming a more important source of sediment trap material in late summer and fall.

Information provided by examining these markers may be used to draw conclusions about carbon cycling and transfer of material through the food web. Plankton samples, containing fresh material, provide information concerning fatty acid sources over a very short time frame, allowing the development of the spring bloom to be monitored. Sediment traps provide a more integrated approach, collecting material over a period of several days to months. In the Trinity Bay study, fluxes of fatty acids through the water column were higher during the spring bloom and the traps predominantly captured diatom material at that time. In late summer and fall, terrestrial and zooplankton markers be-

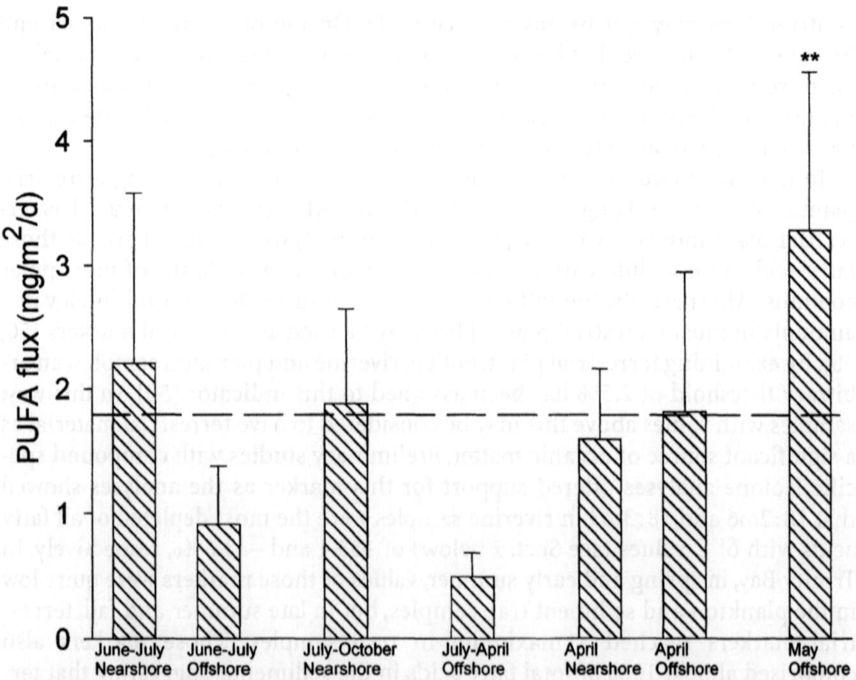

Fig. 4. Fluxes of particulate matter through the water column in Trinity Bay. Rates are the averages of traps at 3 depths (50, 75, 100 m) for polyunsaturated fatty acid inputs: n = 9 – 11. *Error bars* indicate one standard deviation above the mean for each flux; the *dashed line* is the annual mean flux. $^{**}p < 0.05$: Significantly higher than the annual mean flux

came important in the traps. Proportions of PUFA in both plankton and trap samples were high, indicating their efficient transfer to the sea floor. They are essential nutrients for animal survival and growth because they are necessary for normal membrane structure and function, especially at low temperatures [70]. Newfoundland is located in a subpolar oceanographic climate zone, and high levels of PUFA have been found in membrane and storage lipids in benthic organisms living in these cold waters [71].

In this area, there was a high flux of PUFA to the sea floor (Fig. 4) but there was very little preservation in sediments, indicating that the lower food web in this environment had an ample supply of lipids of high nutritional value and that these nutrients are very efficiently recycled. The efficient functioning of the lower trophic levels appears to have been the case for at least the past century as well, as marine lipids and fatty acids remained at about the same low level throughout a 30-cm core from the centre of Trinity Bay. This suggests that the decline in groundfish stocks in this area over the past three decades cannot be related to major shifts in the supply of energy or essential nutrients to the food web. However, the sink for all the marine lipid material sedimenting out of the water column remains to be established, since it is not the sediments and there has been a serious decline in groundfish in the area.

4
Sterols

Sterols are also potentially excellent biomarker compounds due to their stability and the diversity of their structures. They are present in all eukaryotes, and in marine material such as sediments, detection of 25 sterols or more is common. They share with phospholipids a structural function in membranes where, because of their unique hydrophobic and steric properties, they act as specific internal regulators of membrane fluidity and influence various membrane functions and membrane associated enzymes [72].

To determine sterols, a portion of the total lipid extract is saponified using methanolic potassium hydroxide, and sterols subsequently recovered in 2:1 hexane/chloroform. The sterols are converted to the corresponding trimethylsilyl (TMS) ethers using bis-*N,O*-(trimethylsilyl)trifluoroacetamide, BSTFA, and analyzed by capillary GC and GC with mass spectrometry. Reviews of relative retention times and mass spectra for sterol TMS ethers have been published [e.g. 73]. In some cases, sterol acetates, rather than TMS ethers, are the derivatives prepared for GC. Silica column chromatography of the total lipid extract may also be used instead of saponification to isolate the sterol fraction [74], or even sterol subclasses such as 4,4-dimethyl, 4-monomethyl and 4-desmethyl sterols [75], prior to derivatization. However, this approach only includes free sterols in the analysis, whereas by saponifying the total extract, sterols present as steryl esters are also detected.

In algae and in many invertebrates that feed directly on algae there are a wide variety of sterols that can be used for chemotaxonomic purposes and for food web tracing [76–79]. Their comparative resistance to degradation makes them even more valuable as long-term biomarkers than the lipid classes and fatty acids discussed above. Also they provide less ambiguous markers of terrestrial plants, phytoplankton, macroalgae, and of human sewage [e.g. 74]. For example, fresh domestic wastewater can be identified by high coprostanol/cholesterol and 24-ethylcoprostanol/β-sitosterol ratios [80]. 24-Methylenecholesterol can be used as a marker for diatoms [3], and 24-ethylcholesterol, ethylcholest-5,22-dienol and 24-methylcholesterol for terrestrial plants [81], although the use of C_{29} sterols as higher plant markers requires caution, since certain algae also synthesize them [75]. Lastly, the ratio of corresponding saturated to Δ^5-unsaturated species (stanol/stenol ratio) can be used to indicate the preservational state of marine material. The use of supporting evidence from other biomarkers or from floristic analyses greatly increases the certainty of sterol source assignment.

In our Trinity Bay study we found that the input of anthropogenic hydrocarbons was generally low and that the same was true of human sewage which could be ascertained from the sterol analyses. No coprostanol was detected at offshore sites. Very low levels of coprostanol may be present in sediments near towns, but its ratio to cholesterol and epicoprostanol make a sewage source unlikely. By contrast, coprostanol is the principal sterol (20 μg/g dry weight sediment) in harbour sediments in St. John's, the major city on the island, where untreated sewage is also discharged. These results indicate that sewage input in the Trinity Bay area is either negligible, or is being efficiently degraded or dispersed.

Sterol biomarkers can be used to apportion inputs to ecosystems (Fig. 5), with specific sterols being assigned to different source organisms or categories (see also [82]). In sediments from our Trinity Bay study, total sterol concentrations in sediments averaged 24–44 µg/g dry weight, with no discernible trend with depth, suggesting that sterols are well preserved. Source assignment for cholesterol and cholestanol, widespread in marine organisms, was done as follows. In sediments a quarter of the cholesterol and cholestanol contributions were assigned to each of diatoms, dinoflagellates, zooplankton and macroalgae. In settling particles and plankton, in which no typical dinoflagellate sterols were detected, cholesterol and cholestanol were divided among the other source categories. 5β-Stanols are grouped separately; while not all 5β-stanols are unambiguous sewage indicators, they are all formed by bacterial biohydrogenation.

The C_{27} sterols (desmosterol, cholesterol) and C_{28} sterols (24-methylenecholesterol, 24-methylcholesta-5,22-dienol) predominated in the sterol composi-

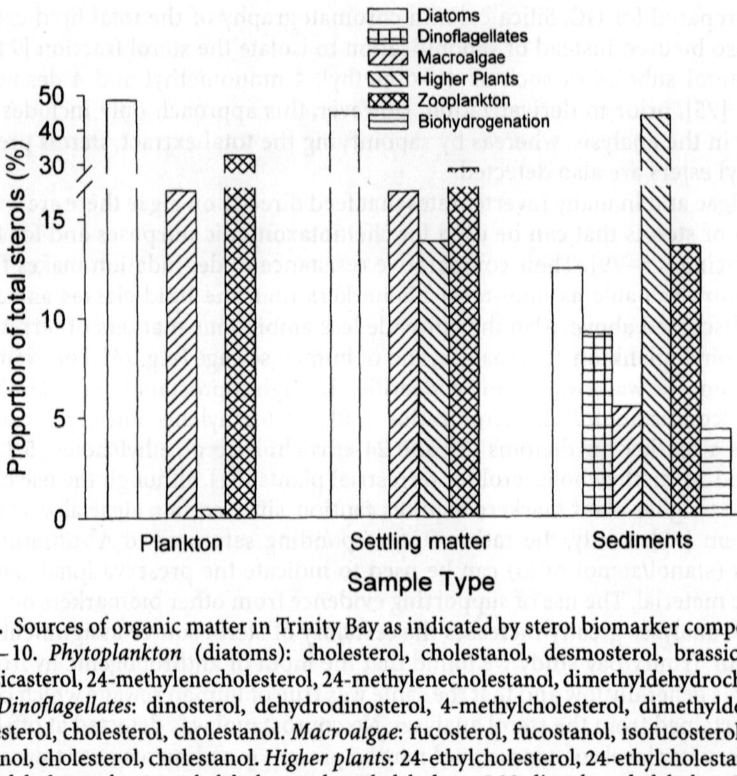

Fig. 5. Sources of organic matter in Trinity Bay as indicated by sterol biomarker composition. n = 8–10. *Phytoplankton* (diatoms): cholesterol, cholestanol, desmosterol, brassicasterol, brassicasterol, 24-methylenecholesterol, 24-methylenecholestanol, dimethyldehydrocholestanol. *Dinoflagellates*: dinosterol, dehydrodinosterol, 4-methylcholesterol, dimethyldehydrocholesterol, cholesterol, cholestanol. *Macroalgae*: fucosterol, fucostanol, isofucosterol, isofucostanol, cholesterol, cholestanol. *Higher plants*: 24-ethylcholesterol, 24-ethylcholestanol, 24-methylcholesterol, 4-methylcholestanol, ethylcholesta-5,22-dienol, ethylcholest-22-enol, 24-ethylcholestanol, $C_{30}\Delta^5$ steratrienol. *Zooplankton*: *trans*-22-dehydrocholesterol, *trans*-22-dehydrocholestanol, 24-nordehydrocholesterol, 24-nordehydrocholestanol, cholesterol, cholestanol, occelasterol. *Biohydrogenation*: Coprostanol, epicoprostanol, ethylcoprostanol. Sterols due to biohydrogenation include both sewage markers and those formed by bacterial degradation

tion of net-tow and sediment trap samples, confirming their mainly marine origin [75]. The diatom sterols 24-methylcholesta-5,22- dien-3β-ol, 24-methylenecholesterol and desmosterol are prevalent in spring/summer sediment trap material, which contained smaller proportions of terrestrial sterols. However, in sediments, C_{29} sterols (e.g. 24-ethylcholesterol) associated with higher plants prevailed (Fig. 5). This indicates a much greater terrestrial contribution, even offshore, and indicates the potential for onshore activities to impact the marine environment. The dinoflagellate contribution (C_{30} sterols) appears to have been well preserved in the sediments. No such contribution was detected in net-tow and sediment trap samples (Fig. 5), probably because those analysed for sterols were taken mainly in spring, while dinoflagellate numbers and biomass proportions increased in summer [50, 83].

5 Hydrocarbons

The last class of lipids being considered here are the hydrocarbons. Hydrocarbon markers include alkanes derived from algae or plant leaves, and polycyclic aromatic hydrocarbons (PAH) derived mainly from crude petroleum and fuel spills. PAH can also be found in combustion products of fuels such as heating oil, gasoline and wood. Coastal sediments act as the ultimate reservoirs for these compounds when they are transported unaltered through the water column.

While hydrocarbons represent only a small fraction of the organic matter present in marine sediments, they have proven to be a class of markers easily extracted by organic solvent (e.g. Soxhlet extraction with dichloromethane) and analysed (GC-MS or GC-FID). Hydrocarbons act as suitable markers for distinguishing different source inputs in marine sediments [84, 85] and for investigating the cycling of organic matter in the marine environment [86–89]. By adopting the approach of studying different classes of hydrocarbons simultaneously (*n*-alkanes, branched alkanes, aromatics etc.) stronger conclusions about carbon sources can be drawn [81, 90].

Certain hydrocarbons are recognized as hazardous environmental compounds. PAHs have been classified as "priority pollutants" because of their carcinogenic and mutagenic characteristics [91]. The health risk associated with PAH, together with the information they offer as environmental markers, justifies inclusion of these compounds in a general study of naturally occurring hydrocarbons (e.g. *n*-alkanes).

In our Trinity Bay study, cores taken from Hickman's Harbour and near Clarenville indicate significant terrestrial inputs which have changed over time [92]. In Hickman's Harbour there has been an increase in the ΣnC_{26}-$nC_{35}/\Sigma nC_{16}$-nC_{25} ratio (nC_{26} represents a straight chain saturated hydrocarbon with 26 carbons; ΣnC_{26}-nC_{35}, sum of abundance of n-C_{26} to n-C_{35} hydrocarbons) during the past century (Fig. 6a) indicating greater leaf wax inputs [87] which may relate to wood cutting. ^{210}Pb dating of a core near Clarenville indicated a much faster sedimentation rate. Here, the ΣnC_{26}-$nC_{35}/\Sigma nC_{16}$-nC_{25} ratio (Fig. 7a) indicates increased deposition of organic compounds from leaves since the 1950s. However,

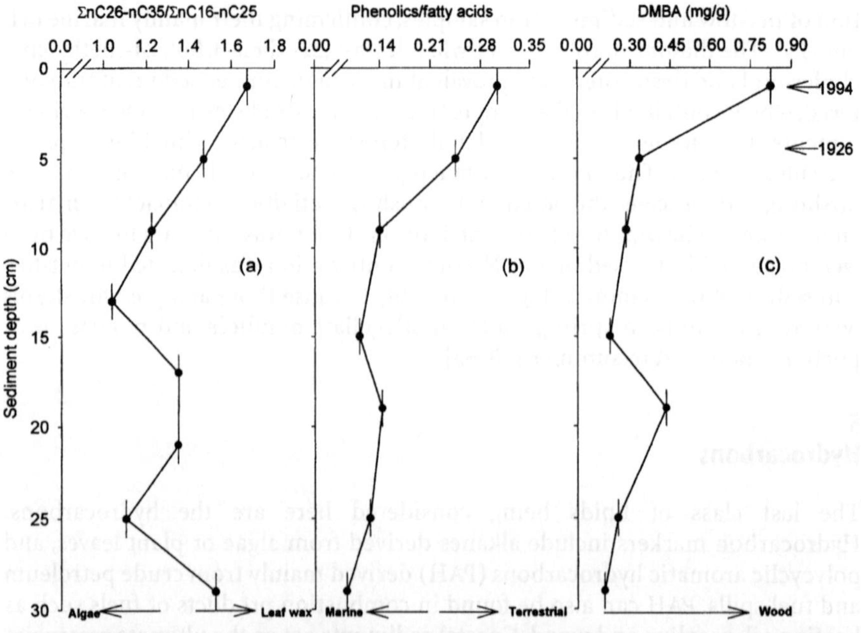

Fig. 6a–c. Biomarkers in a Hickman's Harbour core. a High molecular weight/low molecular weight *n*-alkanes, b phenolic/saturated+branched fatty acid ratio determined by TMAH thermochemolysis, c 3,4-dimethoxybenzoic acid methyl ester (DMBA). ^{210}Pb dates dates were obtained for the top 6 cm of the core. (After [92, 106])

since the mid-1980s there may have been a decline in this source. The total hydrocarbon profile and the high molecular weight/low molecular weight *n*-alkane profiles in the two cores were very similar suggesting the major hydrocarbon source in the area is related to terrestrial plants.

Given the amount of oil field development around Newfoundland, it is critical to establish baseline PAH levels. The input of PAH in this area is apparently low currently except in the vicinity of Hickman's Harbour, where it may be significant. Low levels of various PAH pollutants were identified and their concentrations summed [92]. Total PAH in the Trinity Bay area averaged 0.1 µg/g which is very much lower than St. John's Harbour where a value of 17 µg/g was obtained [93]. St. John's is the biggest city in the Province. The concentration of PAH in Hickman's Harbour surface sediments (0.6 µg/g) was notably higher than in the other Trinity Bay cores suggesting some channeling from the watershed. Among the compounds we examined, the major organic pollutants were the products of wood burning which can be transported mainly through the atmosphere and then deposited by precipitation. This is expected given the amount of wood burning for domestic fuel in the area. The major pyrolytic influences were indicated by the dominance of parental PAH over alkylated PAH [94], and the prevalence in the samples of fluoranthene:pyrene ratios greater than 1 [92]. These inferences have been clearly documented in other studies [81, 89]. Finally, the close resemblance of PAH profiles, together with the similarity

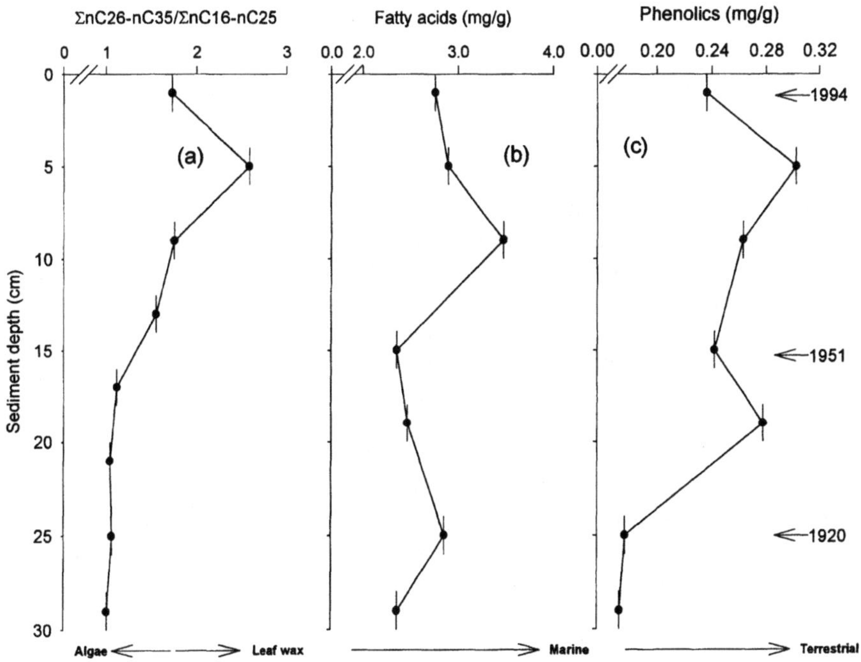

Fig. 7 a–c. Profiles of organic classes in a core taken near Clarenville. **a** High molecular weight/low molecular weight *n*-alkanes, **b** total saturated+branched fatty acids, and **c** total phenolics determined by TMAH thermochemolysis. ^{210}Pb dates were obtained for the top 26 cm of the core. (After [92, 106])

in the phenanthrene/alkylated phenanthrene ratio in all samples, indicate PAH have the same source for the whole area.

6 Pyrolysis Products

Pyrolysis products used as markers include the phenolics. Analytical pyrolysis is a small scale, whole sample, analytical technique which involves the thermal fragmentation of complex organic macromolecules into a range of smaller molecules (pyrolysis products) by rapid application of heat in the absence of oxygen. The products are then identified and quantified by GC-MS to determine the composition of the original sample. The technique can be particularly useful for characterization of complex, non-volatile bio- and geopolymers present in soils, forest humus and aquatic sediments. It is known that simple pyrolysis releases many compounds having phenolic and carboxylic groups which are difficult to analyze by GC. A new technique recently reported for flash pyrolysis with in situ derivatization using tetramethylammonium hydroxide (TMAH) [95] has been shown to be a thermally assisted chemolysis rather than pyrolysis (also called TMAH thermochemolysis) [96, 97]. TMAH not only methylates polar pyrolysis products but also assists in bond cleavage [96]. The technique is now widely used in studying lignin in wood [98, 99] and humic substances in soils [100, 101] and has now been extended to marine sediments [102]. We have adopted the off-line sealed tube technique of McKinney et al. [99] which uses a lower chemolysis temperature (300 °C for 30 min). After excess TMAH has been removed, the products can be extracted with methylene chloride and analyzed by GC-MS [102].

One side reaction which occurs with TMAH/ heat is the formation of benzenecarboxylic acid methyl esters from lignin [103] and from aromatic aldehydes [104, 105]. Therefore our study summed all phenolic-containing products and their methylated derivatives (or "phenolics") observed in marine sediments (Figs. 6–8). Phenolic compounds are abundant in herbs, shrubs and trees. Phenolics can be released during degradation of lignin, from decomposition of leaves and are common plant metabolites.

Lignin is the connective tissue found in plants that gives them rigidity: it comprises 25–30% of the wood of trees and is a generic name for complex irregular phenolic polymers. Lignin is classified into three main groups on the basis of their structural monomer units. One unit common in both hardwood and softwood is coniferyl alcohol. TMAH thermochemolysis products of this unit include 3,4-dimethoxybenzoic acid methyl ester (DMBA), and 3,4-dimethoxybenzaldehyde [106]. DMBA's precursor, vanillic acid, has been used as a lignin-derived marker in marine sediments on the outer Great Barrier Reef

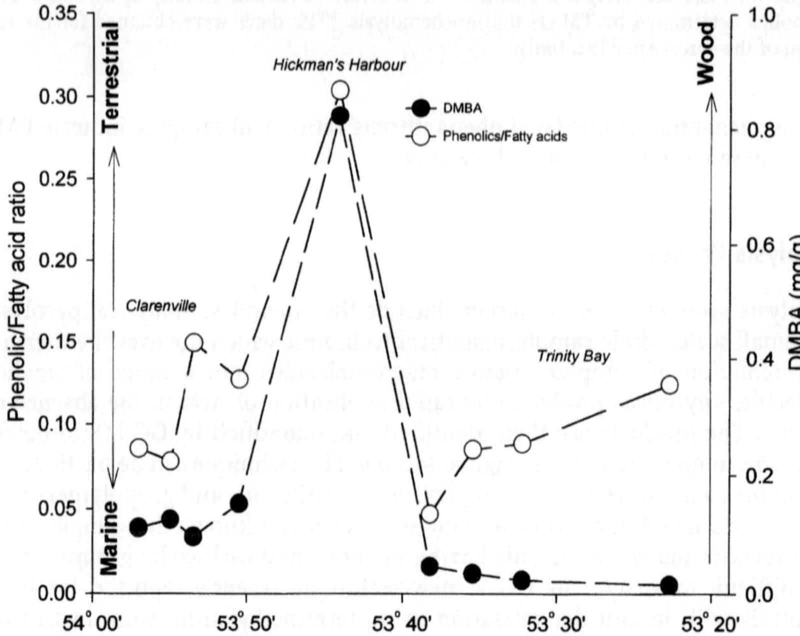

Fig. 8. Ratio of total saturated+branched fatty acids and total phenolics and 3,4-dimethoxybenzoic acid methyl ester content of surface sediments determined by TMAH thermochemolysis. (After [106])

[107]. Pulchan [106] thus used DMBA as a derived marker of vascular plant material (trees and scrubs) in offshore sediments in Trinity Bay (Figs. 6–8). Such terrestrial organic matter may play a significant role in the productivity of marine ecosystems by adding to the marine carbon pool.

Fatty acid methyl esters are another abundant group of products produced by TMAH thermochemolysis. Under these conditions fatty acids of triacylglycerols and other lipids are effectively esterified [97]. But only the saturated/branched fatty acids can be used as reliable biomarkers with this analysis. Unsaturated fatty acids tend to isomerize and degrade under the strong basic conditions using TMAH and heat [108].

In our Trinity Bay study, the TMAH thermochemolysis products as well as the alkane distribution indicated significant terrestrial inputs. The thermochemolysis products included phenolic and saturated/branched fatty acids which were further characterized for their stable carbon isotope composition [102, 106]. The results suggested that the fatty acids were of marine origin while the phenols were derived from terrestrial plants. Although total fatty acid and total phenolic concentrations both increased near Hickman's Harbour, the ratio of their concentrations clearly shows the influence of terrestrial plant inputs there (Fig. 8). The 3,4-dimethoxybenzoic acid methyl ester (DMBA) data further define the source as being derived from terrestrial plants (i.e. the wood and leaves of trees or scrubs). There is five times the concentration of this marker near shore than in Trinity Bay demonstrating the greater influence of wood inputs near the land-margin. However, the widespread occurrence of significant levels of terrestrial markers in the marine sediments indicates that land-derived contaminants could easily become widely dispersed in this system.

Thermochemolysis data from cores from Hickman's Harbour and adjacent to Clarenville [106] also indicate that the degree and type of terrestrial input have changed noticeably over the past century. High concentrations of total phenolics, total fatty acids and total hydrocarbons were found near the surface at Hickman's Harbour and to a smaller extent near the middle of the core. The increase in total fatty acids found near the surface at Hickman's Harbour is smaller than the increase in total phenolics so that the phenolic/fatty acid ratio also increases near the surface (Fig. 6b). The lignin marker (DMBA) and the *n*-alkane ratio point to the significant influence of terrestrial plants, specifically leaf and wood inputs. The higher concentration of marine-derived saturated fatty acids is probably a result of higher preservation in these organic rich sediments. The higher wood contribution to the shallower sediments around Random Island in comparison with the last couple of centuries, suggests a recent change in the nature of terrestrial inputs. This could be a consequence of the natural evolution of the surrounding terrestrial ecosystem, or of human activity such as logging and milling. In 1911 there were 14 sawmills in Hickman's Harbour. The increase in all the organic classes and the biomarkers near the middle of the core is interesting as it likely predates the discovery of Hickman's Harbour by Europeans. However, Hickman's Harbour was probably an area of Indian encampment, and in fact, was one of the last homes of the Beothuk Indians on the east coast.

Due to faster sedimentation rates near Clarenville we have many more samples dating from the last century, and the marked increase in organic compounds derived from terrestrial plants in the 1920s is even clearer as indicated by an increase in phenolics at 20 cm core depth (Fig. 7c). The marked increase in the ΣnC_{26}-nC_{35}/ΣnC_{16}-nC_{25} ratio (Fig. 7a) since the 1950s suggests greater wood cutting for fuel since then. The fact that the profile of the marine-derived fatty acids is completely different to that of the phenolics or of the hydrocarbons, suggests that levels of land use changes have, at most only weakly impacted the pattern of marine biogenic productivity in Trinity Bay.

7 Carbon Isotope Chemistry of Biogenic Compounds

The isotopic composition of organic compounds has been used to speculate on sources of sedimentary organic matter, and on a wide range of paleo-environmental conditions, including temperature, water column stratification, and dissolved inorganic carbon (DIC) concentrations [e.g. 109–112]. The organic matter that is eventually stored in sediments, however, was originally synthesized by various organisms at different trophic levels using a number of distinct biochemical pathways. These differences, as well as factors such as temperature, growth rate, and DIC concentrations contribute to the considerable variation in isotopic composition observed in organic matter. Our limited understanding of these processes, as well as of the distribution, cycling and degradation of organic compounds, means that studies of the isotopic composition of modern environments is important for correct interpretation of fossil isotopic signatures.

Our studies of cold coastal ecosystems of Newfoundland have combined molecular biomarker characterization with the determination of carbon isotope ratios of individual compounds during spring blooms. These studies were conducted to elucidate the transfer of primary photosynthate in the water column to benthic environments. Such studies were also undertaken to measure the range of fatty acid isotopic compositions in a modern depositional environment, and to test the use of isotopic compositions as tracers of fatty acids in settling particles. Compound-specific carbon isotope determinations were made on fatty acids as well as biogenic hydrocarbons (alkanes, highly branched isoprenoids: [113]). Compound-specific carbon isotope measurement can be performed on fatty acids esterified using BF_3-methanol. Compound-specific GC-combustion-isotope ratio mass spectrometry (GC/C/IRMS) analyses of fatty acid methyl esters can be undertaken using a VG Isochrom system equipped with a gas chromatograph. Standardization is accomplished by comparing integrated $^{13}C/^{12}C$ for each compound peak with similar ratios from pulses of reference CO_2 gas introduced before and after the sample chromatographic window. The accuracy of this procedure is tested by co-injection of fatty acid carbon isotopic standards. Measured carbon isotope compositions for esters can be corrected according the procedure outlined by Abrajano et al. [114], and reported as $\delta^{13}C$:

$$\delta^{13}C_{sample} = 1000 * \{(^{13}C/^{12}C_{sample} / ^{13}C/^{12}C_{PDB}) - 1\}$$

$\delta^{13}C$ values, in parts per thousand, are used to describe the small variations in the relative isotope abundances. The key step in the isotopic segregation of carbon is at the point of fixation into plants. For both terrestrial and aquatic plants, photosynthetic fractionation of carbon isotopes arises from differential diffusion of isotopic species of CO_2, H_2CO_3 or HCO_3^-, and from catalytic reactions that vary depending on the fixation pathway. Most marine phytoplankton utilize the C_3 fixation pathway (Calvin-Benson cycle), resulting in bulk $\delta^{13}C$ values in the range –20 to –30‰. The latter value includes the compounding effect of diffusive fractionation and various environmental parameters such as DIC concentrations. The carbon isotopic composition of individual compounds in photosynthesizers also depends on fractionation steps at major branching points of compound synthesis. For example, it has long been known that lipids are generally depleted in ^{13}C compared to bulk $\delta^{13}C$ values of organisms [e.g. 115]. Carbon isotope fractionation in fatty acids, in particular, could occur during the formation of acetyl-CoA and subsequent chain elongation.

Bulk carbon isotope values recorded across spring diatom increases in the cold coastal environments of Newfoundland were reported by Ostrom et al. [116] for Conception Bay (Fig. 1, inset) which is an environment very similar to nearby Trinity Bay. They noted carbon isotope shifts that are consistent with lowered DIC concentrations at the peak of the bloom. This depleted DIC concentration can lead to ^{13}C-enriched organic matter because lower DIC availability also leads to lower isotopic discrimination (i.e. lower isotopic selectivity). Our work focused on compound-specific measurements on individual hydrocarbons and fatty acids in Conception Bay. Bieger et al. [113] noted carbon isotope variations for alkanes and highly branched isoprenoids (HBI) that are consistent with the variations noted by Ostrom et al. [116]. Hydrocarbons generated at the peak of the bloom tend to be enriched in ^{13}C, again suggesting lower isotopic discrimination when DIC substrate concentration is low.

The work of Bieger et al. [113] provided the additional insight that different compounds may record generation at different stages of the bloom. Although this is expected on physiological and biochemical grounds, the work of Bieger et al. [113] was the first to document this isotopically. Most algal products found in plankton and sediments, such as heneicosahexaene (HEH) and pristane, had isotopic compositions between –25 and –28‰. The HBI alkenes, however, were consistently depleted by at least 2‰ (mean $\delta^{13}C = -33$‰) relative to most other marine biogenic compounds. Among the four pairs of HBI alkene isomers, the later eluting isomer was, in each pair, consistently enriched in ^{13}C. The C_{20-25} alkenes found in the near-shore sediments were all significantly enriched in ^{13}C (mean $\delta^{13}C = -20.3$‰) relative both to co-occurring hydrocarbons and to the C_{25} HBI alkenes in the mid-bay samples. The HBI alkenes in crab and scallop samples were isotopically similar to the same compounds in plankton tows and sediments. Squalene was relatively enriched (mean $\delta^{13}C = -24$‰), whereas the C_{25} HBI alkenes were all strongly depleted (from –30.6 to –40.5‰). The average $\delta^{13}C$ of the *n*-alkanes in spring plankton fell over the course of the bloom from –23.7‰ to –29.6‰. The long-chain *n*-alkanes ($>C_{25}$) found in all sediment samples were consistently depleted in ^{13}C ($\delta^{13}C < -30$‰) compared with shorter chain-length homologues. Bieger et al. [113] speculated that the ^{13}C-depleted highly branched isoprenoids were generated prior to the bloom, possibly largely utilizing respired carbon in deeper portions of Conception Bay. While compound-specific carbon isotope variations in alkanes and other biogenic compounds appear to track the effects of decreasing carbon substrate across the spring bloom (i.e. ^{13}C enrichment at the peak of the bloom), there was also isotopic evidence that the short and long chain *n*-alkanes were derived from isotopically unrelated sources.

Our most recent work on fatty acids in Conception Bay [117] has provided additional details to the overall observations made for bulk carbon and individual hydrocarbons discussed above. We combined detailed molecular characterization with carbon isotopic measurements to describe both the temporal and depth variations of primary production during the spring bloom. The isotopic compositions of the fatty acids are summarized in Fig. 9. Bulk spring bloom particulate organic matter in Conception Bay has a $\delta^{13}C$ value between −24‰ and −26‰ [116]. Since lipids are normally expected to be depleted relative to

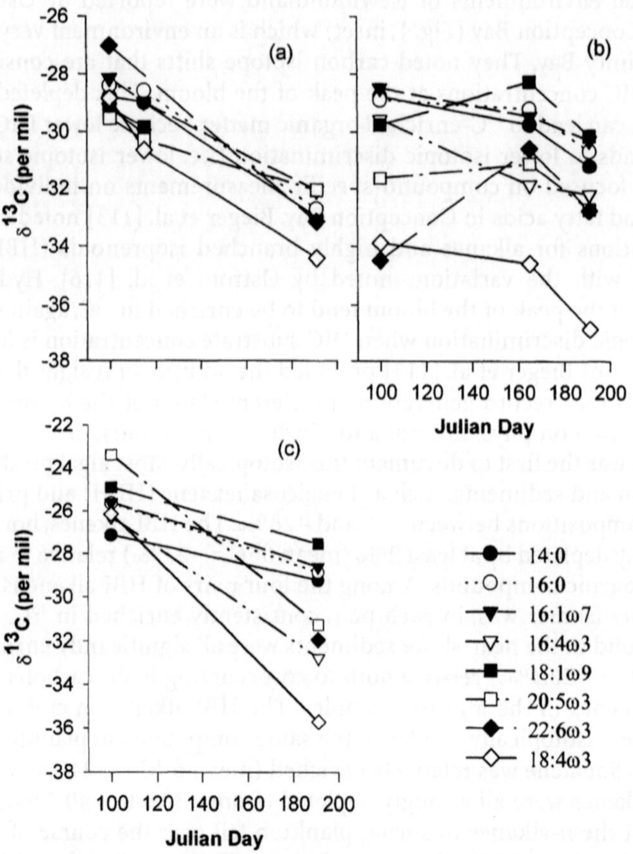

Fig. 9a–c. Carbon isotope composition of individual fatty acids measured across the 1996 spring bloom in Conception Bay, Newfoundland. **a** Horizontal net-tow **b** 80 m depth, $HgCl_2$-poisoned sediment trap and **c** 220 m depth, $HgCl_2$-poisoned sediment trap. (After [117])

total biomass by roughly 5‰ as a result of fractionation during early synthetic stages [115], the isotopic compositions of fatty acids in this study are generally consistent with the expected range for local phytoplankton blooms. A notable exception to this observation is the $\delta^{13}C$ values recorded for 18:4ω3 especially at the waning stages of the bloom. This observation was first noted in Trinity Bay mussels by Abrajano et al. [114], who suggested substantial fractionation during desaturation or a possible additional source for 18:4ω3. Note that the compound-specific $\delta^{13}C$ values for fatty acids overlaps with those previously reported by Abrajano et al. [114] for Trinity Bay mussels. In general, unsaturated fatty acids are depleted in ^{13}C compared to saturated counterparts, but this is most apparent in samples collected at the waning stages of the bloom. Here, the ω3 fatty acids are clearly systematically depleted in ^{13}C compared to the saturated counterparts, and this was most evident for 18:4ω3. Other $\delta^{13}C$ observations made on Trinity Bay mussels [114] also showed consistent ^{13}C depletion in PUFA, possibly resulting from *de novo* carbon isotope fractionation or additional sources for PUFA (specifically the ω3 fatty acids). Although previous authors have found fatty acid desaturation to be associated with a preferential loss of ^{13}C [118], the lack of a similar isotopic depletion among these highly unsaturated fatty acids suggests that relatively little isotopic discrimination occurs during desaturation.

The durability of the isotopic signatures of fatty acids in the water column of Conception Bay is demonstrated by the consistency of the isotopic compositions in plankton, particulate and benthic macrobiota. It is evident that the absolute $\delta^{13}C$ values and their total range overlap within the analytical variability. More importantly, the intermolecular variations (e.g. depleted ^{13}C in ω3 fatty acids) noted in plankton are largely preserved in the trap materials down to 220 m depth (Fig. 9c). These observations imply that $\delta^{13}C$ individual compound "signatures" of primary producers can be traced into the benthic environment, although sedimentary signatures are a temporal average.

The most noteworthy temporal carbon isotopic trend observed in fatty acids across the spring bloom is the more depleted $\delta^{13}C$ values subsequent to the main phase of the bloom. This observation is consistent with those noted above for biogenic hydrocarbons (e.g., alkanes, HBI) by Bieger et al. [113]. The temporal trend in isotopic compositions was possibly due to either lowered DIC concentrations at the peak of the bloom or very high growth rates of phytoplankton for the same period [113]. As shown by Laws et al. [119], the isotopic composition of primary producers is a function of growth rate, in that the isotopic discrimination factor involved in the assimilation and fixing of DIC is reduced during periods of intense growth. Thus, compounds synthesized before or after a bloom, when growth rates are lower, would be expected to have lower $\delta^{13}C$ values. If the $\delta^{13}C$ values arose during fatty acid synthesis, the clearer $\delta^{13}C$ distinction between saturates and PUFAs observed at the end of the bloom would imply that the depletion in ^{13}C is likely related to growth rate. Changing substrate DIC concentration is more likely to shift $\delta^{13}C$ for individual fatty acids to the same extent.

Compound-specific isotope analyses were also used to help determine the origins of some of the important hydrocarbons and thermochemolysis products isolated from Trinity Bay sediment core sections [92, 102]. Phenols and fatty acids produced by thermochemolysis of sediments were characterized for their stable isotope composition and then compared with samples taken from the topsoil near the marine sites. In general, the results suggested that the fatty acids were of marine origin while the phenolics were derived from lignin.

8 Chemometrics

Chemometrics are the use of mathematical methods in chemistry, such as the use of applied statistics for chemical analyses. They are widely used as a tool in pure research and in industry [120] for the investigation of large amounts of data. Chemometrics are used in experimental design such as the factorial type of experiment where it is determined how the result is affected by simultaneous changes in different factors or variables. The major advantage of the factorial experiment compared with the single-variable-at-a-time method is that it is sensitive to interactions between factors [121]. This multivariate approach has been used to optimize autoinjection procedures and column and injector temperature programs in the gas chromatographic analysis of marine lipids [29].

The application of chemometrics to marine biomarker data is becoming increasingly common as it permits data reduction and an objective interpretation of the results. When there are only a few samples and a few measurements made on each sample then a correlation matrix can be quite useful. For example, Parrish [122] used a correlation matrix to demonstrate the decoupling of dissolved and particulate lipid classes during a spring bloom. This approach can be especially interesting when different types of chemical data are compared or when biological data are compared with chemical data. Derieux et al. [36] found a strong relationship between TAG and 20:5ω3 and between TAG and 22:6ω3 in marine dissolved and particulate matter which they attribute to preferential degradation of PUFA in energy reserves. By correlating fatty acid analyses with biovolumes derived from seston population analysis, Parrish et al. [44] were able to confirm 14:0 and the ratio 16:1/16:0 as markers for diatoms, 18:5ω3 and 20:4ω6 as microzooplankton markers, and 15:0 and 22:6ω3 as cryptophyte markers. A more sophisticated use of correlation coefficients involves the multivariate technique of cluster analysis. This type of analysis is available in software packages such as Minitab and its purpose is to classify observations into groups when the proper grouping is initially unknown. Colombo et al. [123] used cluster analysis on organic classes measured in settling particles in the Laurentian Trough. A dendogram based on the correlation coefficients shows the data falling into a phytoplankton group and a zooplankton group and it suggests that lipids and amino acids are the principal contribution of zooplankton to the vertical flux of carbon.

In the marine biomarker field, principal components analysis (PCA) is gaining in popularity as a powerful data reduction procedure. This multivariate technique handles a large amount of variables at the same time, instead of the traditional pairwise correlation studies. With the use of autoinjectors on gas chromatographs and integration software that can be directly linked with spreadsheets it is possible to quickly generate large amounts of data that would be very difficult to interpret by correlation analysis. In such a situation, a multitude of significant correlations can be generated and it is very difficult to visualize and then utilize the results. PCA, on the other hand, gives a simple graphical representation of the similarities in the data set which allows all the observations (variables) to be considered at the same time. It establishes the principal characteristics of a set of variables on a series of samples by systematically reducing all the original variables to a smaller more coherent set of derived variables (principal components) that capture most of the information contained in the original variables [124]. In the process, it maximizes the variance accounted for in the original variables. It is similar to other procedures such as factor analysis, discriminant analysis, canonical correlation analysis, principal coordinate analysis and correspondence analysis. PCA is more commonly used with lipid data; however, what was probably the first multivariate study of marine lipid data was performed using correspondence analysis [125].

PCA can be easily performed on a data set using statistical software packages such as Minitab and the results displayed using plotting packages such as SigmaPlot (Figs. 10–12). Normality of the data is not required for PCA, however, both Jeffries and Lambert [126] and Mayzaud et al. [127], early users of PCA of lipid data, transformed their percentages. While a whole series of principal components is calculated, often just the first two explain a large proportion (>70%) of the total variance. Thus, by plotting the coefficients of these two principal components a good approximation of the distribution of the observa-

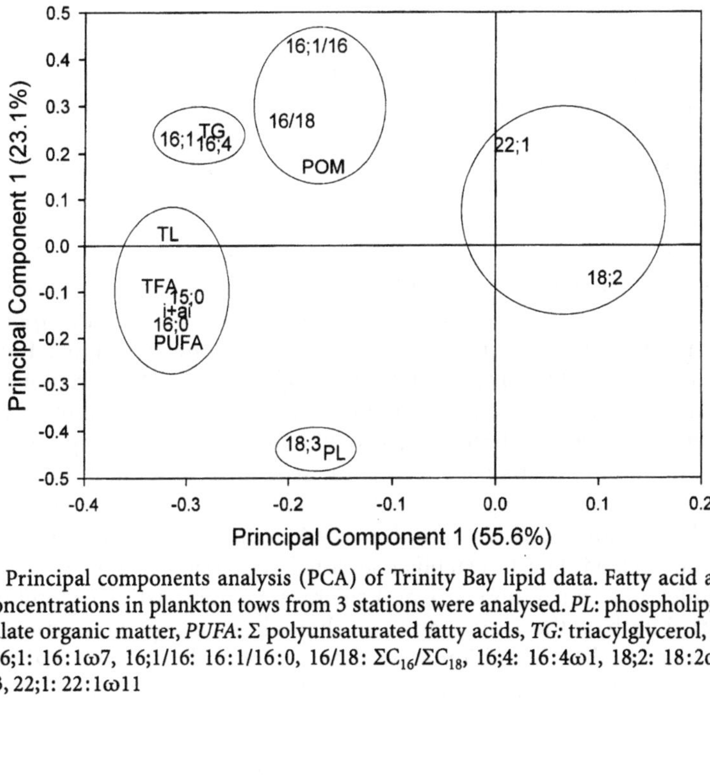

Fig. 10. Principal components analysis (PCA) of Trinity Bay lipid data. Fatty acid and lipid class concentrations in plankton tows from 3 stations were analysed. *PL*: phospholipid, *POM*: particulate organic matter, *PUFA*: Σ polyunsaturated fatty acids, *TG:* triacylglycerol, *TL:* total lipid, 16;1: 16:1ω7, 16;1/16: 16:1/16:0, 16/18: $\Sigma C_{16}/\Sigma C_{18}$, 16;4: 16:4ω1, 18;2: 18:2ω6, 18;3: 18:3ω3, 22;1: 22:1ω11

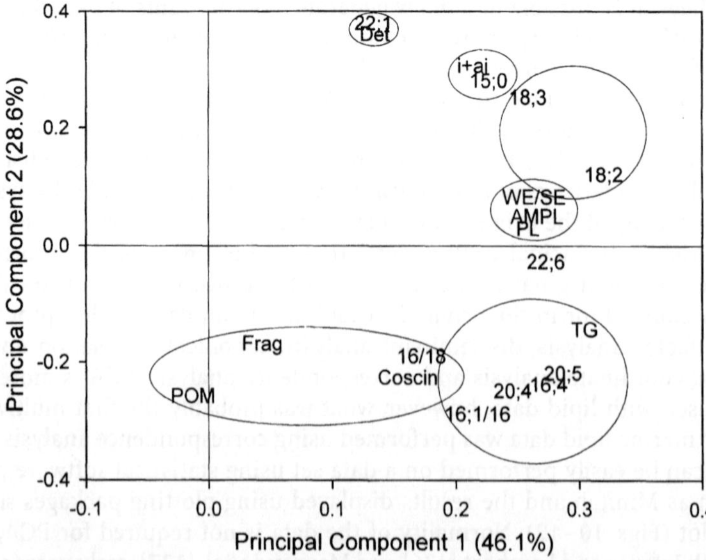

Fig. 11. Principal components analysis of lipid and microscopy data from the water column in Trinity Bay. All data used were concentrations per dry weight. AMPL: acetone-mobile polar lipids, *Coscin*: *Coscinodiscus*, *Det*: detritus, *Frag*: *Fragilaria*, *i+ai*: *iso+anteiso*, *PL*: phospholipid, *POM*: particulate organic matter, *TG*: triacylglycerol, *WE/SE*: wax and steryl esters, 16;1/16: 16:1/16:0, 16/18: $\Sigma C_{16}/\Sigma C_{18}$, 16;4: 16:4ω1, 18;2: 18:2ω6, 18;3: 18:3ω3, 20;4: 20:4ω6, 20;5: 20:5ω3, 22;1: 22:1ω11, 22;6: 22:6ω3

tions is obtained. The plot allows all variables to be considered simultaneously to determine groups and outliers. Groups can be encircled manually [e.g. 126, 127] or rays can be drawn to the variables as used by Colombo et al. [82] in a study of particulate lipids in the Laurentian Trough. There are clustering algorithms that can be used, but Dunteman [124] suggests that visual clustering is as good. We prefer to use a software drawing procedure that limits the group to the shape of a circle or an ellipse that can stretch only along the axes of the principal components (Figs. 10–12). If the first two principal components do not account for a substantial amount of the variation then it is useful to supplement with information from the third one. This could be done by simply adding the sign of the coefficient to each point on the original plot [124], by replotting the coefficients of the first principal component against those of the third principal component [127], or in a three-dimensional plot [125]. If sampling was undertaken in a continuous spatial or temporal sequence then the plotting of the scores of the principal components for each sample can be very useful in addition to plotting the coefficients. In this way Mayzaud et al. [127] used PCA on arcsin transformed data to establish seasonal succession in seston in a small marine inlet, in terms of both size and chemical characteristics. Colombo et al. [82] plotted both on the same graph and showed that a large proportion of the total variance in their lipid biomarker data was related to a terrestrial-marine or vascular plant-phytoplankton gradient in the Laurentian Trough.

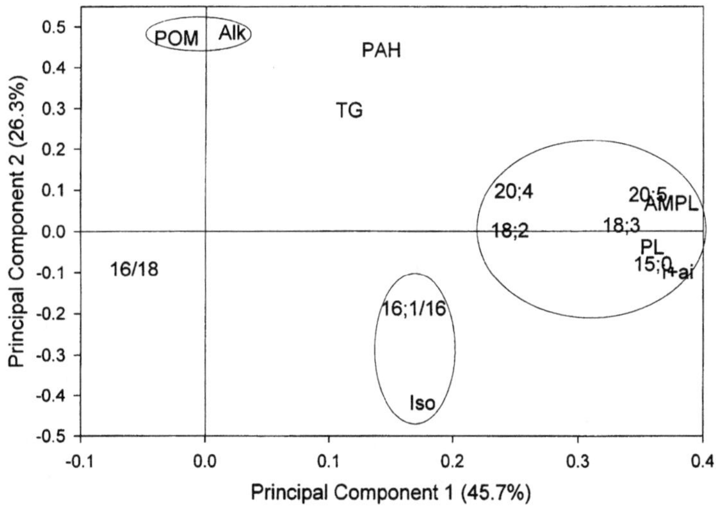

Fig. 12. Principal components analysis of sediment data from Trinity Bay. All data used were concentrations per dry weight. *Alk*: Σ alkanes, *AMPL*: acetone-mobile polar lipids, *i+ai*: *iso+anteiso*, *Iso*: Σ isoprenoid hydrocarbons, *PAH*: Σ polycyclic aromatic hydrocarbons, *PL*: phospholipid, *POM*: particulate organic matter, *TG*: triacylglycerol, 16;1/16: 16:1/16:0, 16/18: $\Sigma C_{16}/\Sigma C_{18}$, 18;2: 18:2ω6, 18;3: 18:3ω3, 20;4: 20:4ω6, 20;5: 20:5ω3

We found very little difference between using percentage data and arcsin transformed data, but it is important to note that there was a substantial difference in the appearance of the plots of the first two principal components depending on whether percentage or concentration data were used. Shown in Figs. 10–12 are PCA of concentration data. In an analysis of lipid class and fatty acid data from three stations in Trinity Bay (Fig. 10) the first two principal components account for a large amount of the variation in the original data (79%). The first axis separates microorganisms (16:1ω7 and PUFA from microalgae and *iso+anteiso* acids and 15:0 from bacteria) and higher organisms (22:1ω11 from zooplankton and 18:2ω6 from terrestrial plants). The second axis shows a weaker separation of plant lipids. Diatom markers generally show a more positive loading on the second axis with 16:1/16:0 showing the most positive loading and terrestrial markers generally show a more negative loading with 18:3ω3 showing the most negative loading. The grouping of the bacterial markers with PUFA and 16:1ω7 suggests that bacteria are associated with phytoplankton. The association of triacylglycerols with 16:1ω7 and 16:4ω1 suggests diatoms are the major source of triacylglycerols and that these acids are significantly associated with storage. The central position of POM shows the multiple contributions but its closeness to diatoms markers signifies their importance as a source in plankton.

PCA of fatty acid and lipid class data for net-tows and traps from various stations consistently showed total lipids and 16:0 grouping together reflecting the ubiquity of this acid in acyl lipids. 15:0 and *iso + anteiso* acids always grouped together confirming that 15:0 has an important bacterial source. In addition, ΣPUFA were also always associated with the bacterial group suggesting bacterial activity was associated with algae. Also 16:4ω1 and TG were usually associated in tows and traps suggesting diatoms are a major source of TG throughout the water column. 16:1/16:0 and $\Sigma C_{16}/\Sigma C_{18}$ were consistently located near organic matter. Similar patterns were found in the traps whether the data were expressed as concentrations or fluxes. On the basis of these similarities and the fact that lipolysis indices were similar in the traps and the tows, the lipid class and fatty acid data from tows and traps were combined and then compared with biovolume data estimated by microscopy (Fig. 11). The appearance of this plot is quite similar to that obtained when PCA is performed on sediment trap data alone.

For the net-tow and sediment trap data combined (Fig. 11) the first 2 PCs account for 75% of the variation in the original data. PC1 separates pennate diatoms (*Fragilaria*) and terrestrial material (18:2ω6 and 18:3ω3) while PC2 separates centric diatoms (*Coscinodiscus*) from detritus and zooplankton material (22:1ω11). The location of the bacterial markers (15:0 and *iso* + *anteiso* acids) suggests they are associated both with terrestrial material and with detritus. Triacylglycerols are again located near the diatom markers. *Coscinodiscus* is also located close to the diatom markers ($\Sigma C_{16}/\Sigma C_{18}$, 16:1/16:0, 16:4ω1) and two PUFA which are quite prominent in the genus [128]. 22:6ω3 which is more prominent in dinoflagellates and flagellates than in diatoms, is separated from other fatty acids. The location of the centric and pennate diatoms near POM underlines the importance of diatoms as overall contributors to organic matter in the water column. Wax and steryl esters, acetone-mobile polar lipids and phospholipids are centrally located indicating the diversity of sources, although wax and steryl esters are located closer to 22:1ω11 and phospholipids to the long-chain PUFA, reflecting their major sources.

Hydrocarbon analyses performed on the sediments from Trinity Bay [92] were included in the analysis of the sediment data (Fig. 12). Here, the first 2 PCs account for 72% of the variation. PC1 separates diatoms ($\Sigma C_{16}/\Sigma C_{18}$) and bacteria (*iso* + *anteiso*) while PC2 separates marine (isoprenoid hydrocarbons) and terrestrial (*n*-alkanes) material. The isoprenoid hydrocarbons consisted mainly of C_{25} highly branched alkenes while the *n*-alkanes were dominated by *n*-C_{27} and *n*-C_{29} [92]. The latter were located with POM signifying the importance of terrestrial plant material in sediments. The diatom marker 16:1/16:0 groups with the isoprenoids indicating these contributions are linked. 15:0 and *iso* + *anteiso* come together suggesting bacteria are a major source of 15:0 in sediments as well. The location of the *n*-alkanes and the C_{18} PUFA suggests different sources for these terrestrial markers. 20:4ω6 is close to 18:2ω6 and 18:3ω3 suggesting a similar terrestrial contribution to 20:4ω6. Total PAH is located near total alkanes indicating a similar pathway of delivery. Triacylglycerol is located nearer to the centre suggesting a multiplicity of sources.

Thermochemolysis [106] and sterol analyses were also performed on other sediment samples from Trinity Bay. In the core data all the terrestrial thermochemolysis markers (total phenolics, total phenolics/total fatty acids and 3,4-dimethoxybenzoic acid methyl ester) consistently grouped together and were always located near POM. In addition, the marine thermochemolysis marker (total fatty acids: [106]) was always associated with dinosterol in sediment data ΣPCA.

Mayzaud et al. [127] also used an elaboration of PCA: PCA on instrumental variables, where one group of variables is selected to explain variations in the other group. In this way they showed close associations between small particles, characteristic of summer, and 16:4ω3, 18:3ω3, 18:4ω3, 18:5ω3, 22:6ω3 and 22-dehydrocholesterol, and between medium-sized particles, characteristic of the spring bloom and 16:0, 16:1ω7, 16:3ω4, 16:4ω1 and 24-methylenecholesterol. In another seasonal study, Galois et al. [43] used correspondence analysis to show an association of a high bacterial biomass with spring blooms and with river detritus, especially in winter. In another area strongly affected by riverine inputs Yunker et al. [81] used PCA to show that 24-ethylcholesterol, ethylcholest-5,22-dienol and 24-methylcholesterol were terrigenously derived.

PCA and related techniques have been used at several trophic levels above the base of the food web. Using PCA on macrozooplankton, Jeffries and Lambert [126] found 18:0 characterized riverine zooplankton. Following direct hydrolysis and derivatization of fish eggs, Vogt et al. [129] used PCA on 24 GC peaks to distinguish between cod and haddock and between different stages for each species. Navarro et al. [130] used discriminant analysis of fatty acid data to distinguish sea bass larvae fed different diets and anatomical differences in responses to different diets. Two PCA studies of fatty acids in seals suggested that fatty acid composition of dietary lipids was significantly altered before deposition in blubber fat [131] complicating the determination of dietary influences, but that it was possible to distinguish different populations using jaw bones [132]. However, multivariate analysis of a much larger data set [133] has shown that fatty acid signatures of prey are, in fact, reflected in blubber fat.

While chemometrics have been successfully applied to marine food web compartments ranging from dissolved matter to seals an important next step is to use multivariate analyses of biomarkers to objectively define and then quantify trophic relationships in marine ecosystems.

9
Conclusions

Lipid biomarkers are being incorporated to a greater and greater extent in ecological studies; however their use in combination with other techniques greatly strengthens their value. At a first level, the use of individual compounds with synoptic class information or else individual compounds from different classes greatly enhances confidence in source identification. Use of lipids in conjunction with non-lipid markers or stable isotopes can bring a completely new dimension to the picture, especially if multivariate analysis is used. In ecological studies, multivariate analyses are particularly strong when biological data are incorporated into the matrix along side molecular data. We used this approach to investigate the natural biological inputs to Trinity Bay together with the human impacts over the past century. We found that marine compounds enter the food web in a very strong seasonal cycle with maximum inputs occurring in spring. At this time biogenic fluxes are very high, as is the quality in terms of energy content and essential fatty acids. Very little of this material is buried in the sediments suggesting that it is sequestered in the food web and that the lower trophic levels are functioning efficiently. This appears to have also been the case in the past. Taken together, these observations indicate that the decline in groundfish stocks in this area over the past three decades cannot be related to major shifts in the supply of energy or essential fatty acids to the food web. However, the sink for all the marine lipid material sedimenting out of the water column remains to be established, since it is not the sediments and since there has been a serious decline in groundfish stocks in the area. This represents our next challenge in marine ecosystem studies. Simply identifying sources will not be enough: we are now going to need to be able to quantitatively apportion sources, to identify trophic pathways and to quantify sinks.

Acknowledgements. We wish to thank the captains and crews of the F.R.V. *Shamook*, the R.V. *Karl and Jackie II*, the M.V. *Nain Banker*, and the *Baccalieu Endeavour*. We also thank Jeanette Wells and Linda Winsor for technical support in analytical aspects of this work, and Yvette Favaro for use of some data from her M.Sc. thesis. This work was funded through Environment Canada's Tri-Council Eco-Research Program.

10
References

1. Villanueva J, Grimalt JO, Cortijo E, Vidal L, Labeyrie L (1997) Geochim Cosmochim Acta 61:4633
2. Guzman-Vega MA, Mello MR (1999) Amer Assoc Petr Geol Bull 83:1068
3. Volkman JK, Barrett SM, Blackburn SI, Mansour MP, Sikes EL, Gelin F (1998) Org Geochem 29:1163
4. Forget J, Pavillon J-F, Beliaeff B, Bocquene G (1999) Env Toxicol Chem 18:912
5. Mora P, Fournier D, Narbonne J-F (1999) Comp Biochem Physiol C 122:353
6. Piechotta G, Lacorn M, Lang T, Kammann U, Simat T, Jenke H-S, Steinhart H (1999) Ecotox Env Safety 42:50
7. Sturm A, da Silva de Assis HC, Hansen P-D (1999) Mar Env Res 47:389
8. Sargent JR, Parkes RJ, Mueller-Harvey I, Henderson RJ (1987) In: Sleigh MA (ed) Microbes in the Sea. Ellis Horwood, Chichester, pp 119–138
9. Saliot A, Laureillard J, Scribe P, Sicre MA (1991) Mar Chem 36:233
10. Conte MH, Eglinton G, Madureira LAS (1995) Phil Trans R Soc Lond B 348:169
11. Hutchings JA, Myers RA (1994) Can J Fish Aquat Sci 51:2126
12. Gomes MC, Haedrich RL, Villagarcia MG (1995) Fish Oceanogr 4:85
13. Folch J, Lees M, Sloane Stanley GH (1957) J Biol Chem 226:497
14. Parrish CC (1988) Mar Chem 23:17
15. Parrish CC (1999) In: Arts MT, Wainman BC (eds) Lipids in freshwater ecosystems. Springer, Berlin Heidelberg New York, pp 4–20
16. Conte MH, Bishop JKB (1988) Lipids 23:493
17. Olsen RE, Henderson RJ (1989) J Exp Mar Biol Ecol 129:189
18. Bergen BJ, Quinn JG, Parrish CC (2000) Env Toxicol Chem: in press
19. Parrish CC, Bodennec G, Gentien P (1992) J Chromatogr 607:97
20. Fraser AJ, Taggart CT (1988) J Chromatogr 439:404
21. Ohman MD (1997) J Plankt Res 19:1235
22. Parrish CC, Ackman RG (1983) Lipids 18:563
23. Rao GA, Riley DE, Larkin EC (1985) Lipids 20:531
24. Sasaki GC, Capuzzo JM (1984) Comp Biochem Physiol 78B:525

25. Fraser AJ, Tocher DR, Sargent JR (1985) J Exp Mar Biol Ecol 88:91
26. Parrish CC (1987) Can J Fish Aquat Sci 44:722
27. Vanderploeg HA, Gardner WS, Parrish CC, Liebig JL, Cavaletto JF (1992) Limnol Oceanogr 37:413
28. Parrish CC, Bodennec G, Gentien P (1996) J Chromatogr A 741:91
29. Yang Z, Parrish CC, Helleur RJ (1996) J Chromatogr Sci 34:556
30. Fraser AJ (1989) Can J Fish Aquat Sci 46:1868
31. Wakeham SG, Lee C, Farrington JW, Gagosian RB (1984) Deep-Sea Res 31:509
32. Parrish CC, Wangersky PJ (1987) Mar Ecol Prog Ser 35:119
33. Parrish CC, Wangersky PJ (1990) J Plankt Res 12:1011
34. Roessler PG (1990) J Phycol 26:393
35. Delbeke K, Teklemariam T, de la Cruz E, Sorgeloos P (1995) Intern J Environ Anal Chem 58:147
36. Derieux S, Fillaux J, Saliot A (1998) Org Geochem 29:1609
37. Goutx M, Gerin C, Bertrand JC (1990) Org Geochem 16:1231
38. Gerin C, Goutx M (1993) J Planar Chromatogr 6:307
39. Conte MH, Eglinton G (1993) Deep-Sea Res I 40:1935
40. Sikes EL, Volkman JK (1993) Geochim Cosmochim Acta 57:1883
41. Parrish CC, Wells JS, Yang Z, Dabinett P (1998) Mar Biol 133:461
42. Dunstan GA, Volkman JK, Barrett SM, Garland CD (1993) J Appl Phycol 5:71
43. Galois R, Richard P, Fricourt B (1996) Estuar Coast Shelf Sci 43:335
44. Parrish CC, McKenzie CH, MacDonald BA, Hatfield EA (1995) Mar Ecol Prog Ser 129:151
45. Berge J-P, Gouygou J-P, Dubacq J-P, Durand P (1995) Phytochem 39:1017
46. Budge SM, Parrish CC (1999) Phytochem: 52:561
47. Weeks A, Conte MH, Harris RP, Bedo A, Bellan I, Burkill PH, Edwards ES, Harbour DS, Kennedy H, Llewellyn C, Mantoura RFC, Morales CE, Pomroy AJ, Turley CM (1993) Deep-Sea Res II 40:347
48. Budge SM, Parrish CC (1999) Mar Chem: submitted
49. Ackman RG (1986) In: Hamilton RJ, Rossel JB (eds) Analysis of Oils and Fats. Elsevier, London, pp 137–206
50. Budge SM, Parrish CC (1998) Org Geochem 29:1547
51. Claustre H, Marty J, Cassiani L, Dagaut J (1988–89) Mar Microbial Food Webs 3:51
52. Bodennec G, Arzul G, Erard-Le Denn E, Gentien P (1994) p 17 in Tests biologiques et chimiques. Edition de l'IFREMER, Direction Environment et Aménagement Littoral, R. INT. DEL. 94.07
53. Nichols PD, Palmisano AC, Smith GA, White DC (1986) Phytochem 25:1649
54. Viso A, Marty J (1993) Prog Lipid Res 32:1521
55. Parkes RJ, Taylor J (1983) Estuarine Coast Shelf Sci 16:173
56. Caudales R, Wells J M (1991) Int J Syst Bacteriol 42:246
57. Harvey HR, Macko SA (1997) Org Geochem 26:531
58. Volkman JK, Johns RB, Gillan FT, Perry GJ (1980) Geochim Cosmochim Acta 44:1133
59. Wakeham SG, Beier JA (1991) Deep-Sea Res 38:S943
60. Haddad RI, Martens CS, Farrington JW (1992) Org Geochem 19:205
61. Harvey HR (1994) Deep-Sea Res 41:783
62. Wakeham SG (1995) Deep-Sea Res I 42:1749
63. Canuel EA, Martens CS (1993) Org Geochem 20:563
64. Santos V, Billett DSM, Rice AL, Wolff GA (1994) Deep-Sea Res 41:787
65. Colombo JC, Silverberg N, Gearing JN (1997) Org Geochem 26:257
66. Napolitano GE, Pollero RJ, Gayoso AM, MacDonald BA, Thompson RJ (1997) Biochem Syst Ecol 25:739
67. Ratnayake WM, Ackman RG (1979) Lipids 14:795
68. Graeve M, Hagen W, Kattner G (1994) Deep-Sea Res I 41:915
69. Albers CS, Kattner G, Hagen W (1996) Mar Chem 55:347
70. Hazel JR, Williams EE, Livermore R, Mozingo N (1991) Lipids 26:277

71. Parrish CC, Yang Z, Lau A, Thompson, RJ (1996) Comp Biochem Physiol 114B:59
72. Wojciechowski ZA (1991) In: Patterson GW, Nes WD (eds) Physiology and biochemistry of sterols. AOCS, Il, p 361
73. Jones GJ, Nichols PD, Shaw PM (1994) In: Goodfellow M, O'Donnell AG (eds) Chemical Methods in Prokaryotic Systematics. Wiley, Chichester, pp 163–195
74. Laureillard J, Saliot A (1993) Mar Chem 43:247
75. Volkman JK (1986) Org Geochem 9:83
76. Patterson GW (1991) In: Patterson GW, Nes WD (eds) Physiology and biochemistry of sterols. AOCS, Il, p 118
77. Teshima S (1991) In: Patterson GW, Nes WD (eds) Physiology and biochemistry of sterols. AOCS, Il, p 229
78. Idler DR, Wiseman P (1971) Int J Biochem 2:516
79. Barrett SM, Volkman JK, Dunstan GA (1995) J Phycol 31:360
80. Quemeneur M, Marty Y (1992) Estuar Coast Shelf Sci 34:347
81. Yunker MB, Macdonald RW, Veltkamp DJ, Cretney WJ (1995) Mar Chem 49:1
82. Colombo JC, Silverberg N, Gearing JN (1996) Org Geochem 25:211
83. Parrish CC (1998) Org Geochem 29:1531
84. Saliot A (1981) Natural hydrocarbons in sea water. In: Dursuma and Dawson (eds) Marine organic chemistry: evolution, composition, interactions and chemistry of organic matter in sea water. Elsevier, New York
85. Bouloubassi I, Saliot A (1991) Fres J Anal Chem 339:765
86. Barrick RC, Hedges JI, Perterson ML (1980) Org Geochem 21:611
87. Farrington JW, Tripp BW (1977) Geochim Cosmochim Acta 41:1627
88. Colombo JC, Pelletier E, Brochu C, Khalil M (1989) Environ Sci Techn 23:888
89. Bouloubassi I, Saliot A (1993) Oceanologica Acta 16:145
90. Lipiatou E, Saliot A (1991) Mar Pollut Bulletin 22:297
91. Jones PW, Leber P (Eds) (1978) Polynuclear aromatic hydrocarbons. Ann Arbor Sci Michigan. 892 p
92. Favaro YL (1998) M.Sc. thesis, Memorial University of Newfoundland
93. O'Malley VP (1994) PhD thesis, Memorial University of Newfoundland
94. Favaro YL, Abrajano TA Jr, Helleur RJ (1996) Proc Biennial International Conference on Chemical Measurement and Monitoring of the Environment. Ottawa, Canada, May 1996.
95. Challinor JM (1989) J Anal Appl Pyrolysis 16:323
96. de Leeuw JW, Baas M (1993) J Anal Appl Pyrolysis 26:175
97. Challinor JM (1991) J Anal Appl Pyrolysis 29:223
98. Clifford DJ, Carson DM, McKinney JM, Hatcher PG (1995) Org Geochem 23:169
99. McKinney DE, Carson DM, Clifford DJ, Minard RD, Hatcher PG (1995) J Anal Appl Pyrolysis 34:41
100. Martin F, del Rio JC, Gonzalez-Vila FJ, Verdejo T (1995) J Anal Appl Pyrolysis 31:75
101. del Rio JC, Gonzalez-Vila FJ, Martin F, Verdejo T (1994) Org Geochem 22:885
102. Pulchan J, Abrajano TA, Helleur R (1997) J Anal Appl Pyrolysis 42:135
103. Hatcher PG, Nanny MA, Minard RD, Dible DM, Carson DM (1995) Org Geochem 23:881
104. Tanczos I, Schoflinger M, Balla J (1997) J Anal Appl Pyrolysis 42:21
105. Tanczos I, Rendl K, Schmidt H (1999) J Anal Appl Pyrolysis 49:319
106. Pulchan K (2000) Ph.D thesis, Memorial University of Newfoundland, in preparation.
107. Susic M, Alongi D (1997) J Chromatogr 758:243
108. Jun-Kai D, Wei J, Tian-Zhi Z, Ming S, Xio-Guang Y, Chui-Chang F (1997) J Anal Appl Pyrolysis 42:1
109. Hayes JM, Freeman KH, Popp BN, Hoham CH (1989) Org Geochem 16:1115
110. Kohnen MEL, Schouten S, Sinninghe Damsté JS, de Leeuw JW, Merritt DA, Hayes JM (1992) Science 256:358
111. Schoell M, McCafferty MA, Fago FJ, Moldowan JM (1992) Geochim Cosmochim Acta 56:1391
112. Schoell M, Schouten S, Sinninghe Damsté J S, de Leeuw J W, Summons RE (1994) Science 263:1122

113. Bieger T, Abrajano TA, Hellou J (1997) Org Geochem 26:207
114. Abrajano TA, Murphy DE, Fang J, Comet P, Brooks JM (1994) Org Geochem 21:611
115. Monson KD, Hayes JM (1980) J Biol Chem 255:11435
116. Ostrom NE, Macko S, Deibel D, Thompson R (1997) Geochim Cosmochim Acta 61:2929
117. Ramos C (2000) Ph.D thesis, University of the Philippines, in preparation
118. Fang J, Abrajano TA, Comet PA, Brooks JM, Sassen R, MacDonald IA (1993) Chem Geol 109:271
119. Laws EA, Popp BN, Bidigare RR, Kennicutt MC, Macko SA (1995) Geochim Cosmochim Acta 59:1131
120. Schonkopf S (1999) American Laboratory April 1999. p 32
121. Adams MJ (1992) In: Haswell SJ (ed) Practical guide to chemometrics. Marcel Dekker, New York, p 181
122. Parrish CC (1987) Mar Ecol Prog Ser 35:129
123. Colombo JC, Silverberg N, Gearing JN (1996) Mar Chem 51:277
124. Dunteman GH (1989) Principal Components Analysis. Sage Publications, Newbury Park, California, p 96
125. Jeffries HP (1979) Am. Nat., 113:643–658
126. Jeffries HP, Lambert RM (1982) pp 91–101 in Estuarine Comparisons, V.S. Kennedy (ed.) Academic Press Inc., New York.
127. Mayzaud P, Chanut JP, Ackman RG (1989) Mar Ecol Prog Ser 56:189
128. Dunstan GA, Volkman JK, Barrett SM, Leroi J-M, Jeffrey SW (1994) Phytochem 35:155
129. Vogt, NB, Moksness, E, Sporstol SP, Knutsen H, Nordenson S, Kolset K (1986) Mar Biol 92:173
130. Navarro JC, McEvoy LA, Amat F, Sargent JR (1995) Mar Biol 124:177
131. Grahl-Nielsen O, Mjaavatten O (1991) Mar Biol 110:59
132. Grahl-Nielsen O, Mjaavatten O, Tvedt E (1993) Can J Fish Aquat Sci 50:1400
133. Iverson SJ, Frost KJ, Lowry LF (1997) Mar Ecol Prog Ser 151:255

Subject Index

MIX
Papier aus verantwortungsvollen Quellen
Paper from responsible sources
FSC® C105338

If you have any concerns about our products,
you can contact us on
ProductSafety@springernature.com

In case Publisher is established outside the EU,
the EU authorized representative is:
Springer Nature Customer Service Center GmbH
Europaplatz 3, 69115 Heidelberg, Germany

Printed by Libri Plureos GmbH
in Hamburg, Germany